Dimensional Analysis

The great principle of similitude

Dimensional Analysis

The great principle of similitude

Jeffrey H Williams

Formerly at Bureau International des Poids et Mesures (BIPM), Sèvres, France

IOP Publishing, Bristol, UK

ISBN 978-0-7503-3655-0 (ebook)
ISBN 978-0-7503-3653-6 (print)
ISBN 978-0-7503-3656-7 (myPrint)
ISBN 978-0-7503-3654-3 (mobi)

DOI 10.1088/978-0-7503-3655-0

Version: 20211201

IOP ebooks

British Library Cataloguing-in-Publication Data: A catalogue record for this book is available from the British Library.

Published by IOP Publishing, wholly owned by The Institute of Physics, London

IOP Publishing, Temple Circus, Temple Way, Bristol, BS1 6HG, UK

US Office: IOP Publishing, Inc., 190 North Independence Mall West, Suite 601, Philadelphia, PA 19106, USA

For Bruce Currie, without whom none of this would have been possible.

Music of the Spheres

For there is a music everywhere there is a harmony, order or proportion; and thus far we may maintain the music of the Spheres; for those well-ordered motions, and regular paces, though they give no sound to the ear, yet to the understanding do they strike a note most full of harmony.

Sir Thomas Browne (Religio Medici; III (9))

Contents

Preface

The purpose of physics is the understanding of the world around us, not the solving of differential equations. Such equations are often a means to an end, but not the end in themselves. And if that end can be reached by simpler means; especially more physically transparent and intuitive means, all to the good.

In this volume, we will look at the technique of dimensional analysis. And how the foundations of dimensional analysis were laid in the first years of the last century; particularly with the work of Lord Rayleigh (John William Strutt, 3rd Baron Rayleigh, 1842–1919). Interestingly, Lord Rayleigh's paper in *Nature* (1915), which demonstrated the power of the technique for handling complex problems without the need for long, difficult experiments, and the solution of coupled differential equations, appeared at about the same time as other applications of dimensional analysis. When presenting his model of the hydrogen atom in 1913, Niels Bohr used a dimensional argument in his article. In addition, in 1911, Albert Einstein looked at the dimensionality of the heat capacity of solids. Evidently, dimensional analysis was very much the *Zeitgeist* in physics at the beginning of the 20th Century.

One of the advantages of treating complex problems by dimensional analysis, rather than by a straightforward solution of a set of equations is that dimensional analysis allows one to incorporate physical insight into the resolution of the problem. This idea goes back to the very first use of dimensional analysis. In 1915, Lord Rayleigh commented that it was a shame that more physicists did not look at problems from the point of view of the underlying dimensions, rather than simply constructing sets of differential equations, and then attempting to solve those equations, or even undertaking a long series of complex experiments. '*I have often been impressed by the scanty attention paid even by original workers in physics to the great principle of similitude. It happens not infrequently that results in the form of 'laws' are put forward as novelties on the basis of elaborate experiments, which might have been predicted a priori after a few minutes' consideration. However, useful verification* [of a hypothesis by experiment] *may be, whether to solve doubts or to exercise students, this seems to be an inversion of the natural order. One reason for the neglect of the principle may be that, at any rate in its applications to particular cases, it does not much interest mathematicians.*' Apart from the splendid Victorian prose, this is quite an indictment from one of the then greatest living physicists. But Lord Rayleigh was making a general comment upon the attitudes of both experimentalists and theoretical physicists, that both should spend more time examining the dimensions involved in the phenomena they investigated.

Dimensional analysis, or the principle of similitude, to use Lord Rayleigh's term, usually gets short shrift in textbooks. What one finds is mostly an admonition about checking the dimensional homogeneity of equations as a way of finding errors. This is good advice, but perhaps lacking in profundity. The remarkable power of dimensional analysis to obtain approximate results quickly and easily is not often fully exploited. By the time most students come across dimensional analysis, they seem to have been thoroughly inculcated in the belief that physics problems must entail the solving of differential equations, and any attempt to side step this rite of passage is a form of cheating. But solving physical problems by dimensional analysis is perfectly respectable; Lord Rayleigh's law involving the inverse fourth power of wavelength for electromagnetic scattering by objects, small compared with the wavelength of the incident radiation, was first obtained by dimensional arguments—and that analysis explains why the sky is blue.

Dr Jeffrey Huw Williams (1956–2021)—an appreciation

I first met the author of this book in September 1967, when a string of accidents brought us from our outlying village schools to the same boys' Grammar school in the centre of the Welsh industrial port of Swansea. It turned out that we had been living a mere two miles apart, but (as ever in Wales) had been separated by one of the ramparts that surround the South Wales Coalfield like a pie crust.

In our second year, we found ourselves in the same class, studying Russian as our second (non-Welsh) modern language. As we waited for our teacher to arrive, Jeffrey suddenly leant over and asked if I knew how atomic bombs worked. It was a strange opener and, happy to plead ignorance, I did. He then explained, with diagrams, the difference between 'Little Boy' and 'Fat Man', and why the latter could deliver bigger yields. And as the teacher arrived, he concluded: 'Next time, I shall explain how a hydrogen bomb works.'

Over half a century later, three things strike me about this. First, his passion to explain, which never left him. Second, that 'next time'. I had been chosen. Third, his abiding interest in the apocalyptic, and in power: where it comes from, what it consists of and how it is wielded. To some extent these things explain his musical tastes—in those days it was Wagner—his interest in the law, in politics, and in religions (despite not being religious himself). With a shared loathing for sport, we discovered a great deal in common. Indeed, only one thing really divided us—admittedly, quite an important thing—and that was algebra.

My shameful inability in that direction means that I shall never be able to fully appreciate this, his final book, which I am happy to have played a small part in rescuing for publication since his sudden death. When he finished it, (earlier than usual, thanks to Covid) he told me, at what proved to be our final meeting, 'I don't think this one's for you, really.' He smiled knowingly. He had days to live.

Like Jeffrey, as a science writer I have also worked for, interviewed and written about many of the world's cleverest people. Yet I can honestly say that I never encountered anyone with a sharper and clearer mind blessed with greater intellectual scope, than the boy from the village over the *Banc* whom I met aged 12; no one, indeed, more rigorous, less sentimental, more clear-headed, or more definite and passionate in his wishes.

Despite his intense interest in history and music, Jeffrey never once doubted he was a scientist; but unlike others who write science for a wider audience (his booklet on HIV, published by the Terence Higgins Trust in 1999, was widely admired and probably saved many lives), Jeffrey's literary destiny was to explain science to fellow scientists. That I feel has been the joy of this series for the Institute of Physics.

He had come to feel that other members of his tribe were perhaps insufficiently open to science's broader implications and interconnections—and that included more than a few of the Nobel Prize winners he encountered in his career. These books seek to address that.

Jeffrey, a lifelong internationalist and Francophile, believed in the big and wide over the small and narrow. He was a breaker of barriers, and a withering critic of the insular. He knew his mind to an extraordinary and enviable degree, and never concealed any aspect of himself from anyone (least of all himself). He always knew what he liked and what he loathed; who he loved, and about whom he need not waste a passing thought. The world is poorer for his loss; but his achievement, in the face of many obstacles, remains an inspiration.

Ted Nield

Dr Ted Nield FGS was Editor, *Geoscientist* magazine and Chair, Association of British Science Writers.

Author biography

Jeffrey Huw Williams

The author enjoying the hot weather in Montpellier, where he has lived for over thirty years.

The author was born in Swansea, UK, in 1956. He attended the University College of Wales, Aberystwyth, and Cambridge University; being awarded a PhD in chemical physics from the University of Cambridge in 1981. Subsequently, his career as a research scientist was in the physical sciences. First, as a research scientist in the universities of Cambridge, Oxford, Harvard and Illinois, and subsequently as an experimental physicist at the Institute Laue-Langevin, Grenoble, which remains one of the world's leading centres for research involving neutrons.

During this research career, the author published more than seventy technical papers and invited review articles in the peer-reviewed literature. However, after much thought the author chose to leave research in 1992 and moved to the world of science publishing, and the communication of science by becoming the European editor for the physical sciences for the AAAS's magazine *Science*.

Subsequently, the author was Assistant Executive Secretary of the International Union of Pure and Applied Chemistry; the agency responsible for the world-wide advancement of chemistry through international collaboration. And most recently, 2003–2008, he was the head of publications and communications at the *Bureau international des poids et mesures* (BIPM), Sèvres. The BIPM is charged by the Metre Convention of 1875 with ensuring world-wide uniformity of measurements, and their traceability to the International System of Units (SI). It was during these years at the BIPM that the author became interested in, and familiar with, the origin of the Metric System, its subsequent evolution into the SI, and the recent transformation into the Quantum-SI.

Since retiring in 2009, the author has devoted himself to writing.

In 2014; *Defining and Measuring Nature: the make of all things*, in the IOP Concise Physics series.

In 2015; *Order from Force: a natural history of the vacuum*, in the IOP Concise Physics series. This title looks at intermolecular forces, but also explores how ordered structures, whether they are galaxies or crystalline solids, arise via the application of a force.

In 2016; *Quantifying Measurement*: *the tyranny of number*, in the IOP Concise Physics series. This title explains the concepts essential in an understanding of the origins of measurement uncertainty. No matter how well an experiment is done, there is always an uncertainty associated with the final result.

In 2017; *Crystal Engineering*: *how molecules build solids*, in the IOP Concise Physics series. This title looks at how the many millions of molecules, of hugely varying shapes and size can all be packed into a handful of crystal symmetries.

In 2018; *The Molecule as Meme*, in the IOP Concise Physics Series. This title explains how the originally separate sciences of physics and chemistry became one science, with the advent of quantum mechanics and the acceptance of the existence of molecules.

In 2019; *The Search for the Absolute*: *How magic became science*, with Morgan & Claypool, San Rafael, CA 94901, USA. This title looks at the origins of science, and the scientific way of looking at the world.

In 2020; the second edition of *Defining and Measuring Nature: the make of all things*, with IOP Publishing, Bristol, UK. This new, and expanded edition details the changes made by the redefinition of four of the seven base quantities of the SI, and the subsequent creation of the Quantum-SI.

In addition, retirement allowed the author to return to the research laboratory, and he began publishing technical papers; this time in the fields of crystal design and structure determination via x-ray diffraction; in particular, the architecture and temperature stability of co-crystals and molecular adducts. The most recent publications are: *Influence of Solvent in Crystal Engineering: A Significant Change to the Order–Disorder Transition in Ferrocene*; Joseph C. Bear, Jeremy K. Cockcroft, and Jeffrey H. Williams; *J. Am. Chem. Soc.* 2020, **142**, 4, 1731–1734, and *Influence of methyl-substitution on the dynamics of the C–H···F–C interaction in binary adducts*; Jeremy K. Cockcroft, Jacqueline G. Y. Li and Jeffrey H. Williams; *CrystEngComm*, 2019, **21**, 5578.

Dr Jeffrey Huw Williams completed this, his last book, on 3 March 2021. Exactly three months later he suffered what proved to be a fatal heart attack at his London home. He never regained consciousness and died, at Royal Trinity Hospice, Clapham Common, on July 9.

Introduction: the language that is science

Definition I: The quantity of matter [mass] *is the measure of the same arising from its density and bulk* [volume] *conjointly...*

Isaac Newton, *Principia* (1687)

A moment's thought will demonstrate that science may be described as the quantitative study of the complex, coupled relationships that may, or may not exist between observed events. Any phenomenon that is susceptible to investigation; that can be measured, that can be weighed, that can be numbered, and that can be expressed mathematically—the readings on laboratory dials, the clicks coming from a counter, or detector may all be considered as part of the enterprise of science. On the other hand, there is no room in the scientific world-view for the inexact, the non-contingent, the immeasurable, the imponderable, or the undefined.

Weights and measures form a part of everyone's ingrained mental landscapes, whether they know it or not. Indeed, it is impossible to function effectively in the modern world without some internalized system of measurement, which enables one to estimate or judge size, volume, weight, duration, length and value. Children may learn something of such systems of weights and measures at school, and then as young adults incorporate them subconsciously for the remainder of their lives. One of the skills that a young scientist will learn to appreciate is the acquisition of a broad physical intuition; that is, being able to: perform 'back-of-the-envelope' calculations or estimations, having experimental skills across different technical areas, and being adept at the use of graphical and statistical data analysis. But above all, young physicists must be able to use mathematics to see the world in quantitative terms.

Dimensional analysis should be familiar to all students in physics; even if only to be able to follow the instruction to 'make sure that the units balance in your final answer', or similar *a posteriori* checks. This volume presents a more *a priori* methodology for the examination, if not the complete solution of many problems involving dimensions. We attempt to show this by first asking (see Lord Rayleigh's comment in the *Preface*) if one can extract the necessary functional dependence of some intermediate, or final answer from the physical parameters relevant to the problem. Having identified such variables, it is often the case that the dimensions of the desired final result can completely determine or, at least, constrain the dependence on those variables—either as being rational powers of combinations of the parameters, and/or as functions of dimensionless combinations.

More fundamentally, we seek to stress that dimensional analysis can be used, almost literally, as a language to describe physical phenomena. The fact that all physical quantities have dimensions that can be written in terms of rational powers of a handful (seven, at present) of base, or fundamental dimensions is a powerful unifying theme in science (these fundamental dimensions are given in table 1.1). We will, in the coming chapters, explore how a basic alphabet of dimensions can be used to encode the description of a huge variety of physical quantities, and they are the means by which new phenomena, as yet unknown to science will be analysed.

Usually, in a physics text designed for a broad readership one begins with one-dimensional kinematics, as only two base dimensions, length (L) and time (T) are required. If $[Q]$ represents the dimensions of quantity, Q; one-dimensional kinematic problems become: $[x(t)] = \mathsf{L}$, $[v(t)] = \mathsf{LT}^{-1}$ and $[a(t)] = \mathsf{LT}^{-2}$, for position, velocity and acceleration, respectively. While some individuals might choose to work in customary units, or units not derived from the International System of Units, the SI (there is nothing wrong with this—all systems of units are but dialects of the single language of science), the use of specific units such as the metres and seconds of the SI does facilitate international communication, as well as communication between the different areas of science. But, in principle, one could just as logically use furlongs for distance and fortnights for duration.

If we extended our study of classical mechanics to include Newton's laws, we would need to introduce the concept of inertial mass, with a new dimension (mass, M); from which some other familiar quantities arise: force, energy (kinetic, potential, relativistic, etc, including work), power, pressure and density (ρ), whose dimensionalities are: $[\text{Force}] = \mathsf{MLT}^{-2}$, $[\text{Energy}] = [\text{Work}] = \mathsf{ML}^2\mathsf{T}^{-2}$, $[\text{Power}] = \mathsf{ML}^2\mathsf{T}^{-3}$, $[\text{Pressure}] = \mathsf{ML}^{-1}\mathsf{T}^{-2}$, and $[\rho] = \mathsf{ML}^{-3}$, respectively. Much of classical mechanics, fluid mechanics, many engineering applications, and even the formalism of quantum mechanics require only the three dimensions, L, T and M[1].

It is through classical mechanics that students encounter the first unification in physics; Isaac Newton's law of universal gravitation. Here the force of gravitational attraction (F) between two masses (m_1 and m_2), separated by a distance, r, is given by $F = G\,(m_1 m_2)/r^2$, where G is a constant of Nature, Newton's gravitational constant. This constant turns the relationship of proportionality between the force of attraction, due to the two separated masses into a relationship of equality. We may determine the units of this constant by looking at the dimensions of the quantities it couples together. On the LHS, we have a force; that is MLT^{-2}, and on the RHS we have $G.(\mathsf{M}^2\mathsf{L}^{-2})$. Consequently, G has dimensions of $\mathsf{L}^3\mathsf{M}^{-1}\mathsf{T}^{-2}$; and today's most precise value of the constant is $6.674\ 30(15) \times 10^{-11}\ \mathrm{m}^3\ \mathrm{kg}^{-1}\ \mathrm{s}^{-2}$. This example of equating dimensions on both sides of an equation is the basis of dimensional analysis.

We note that the principle of equivalence states that inertial and gravitational masses are one and the same, so the same M-dimensionality is used for both. That is, when Newton was sitting in 17th Century Cambridge formulating his ideas of gravity, he was imagining that the concepts he was defining would be as applicable in space, as they were here on planet Earth. This is the Great Principle of Similitude; that is, of similarity.

[1] It is customary that the symbols used to represent the seven base dimensions are represented by capital letters in a *sans-serif* font.

Things become more complex when we introduce electromagnetic quantities: Electric fields are caused by static charges and magnetic fields are caused by moving charges. One can either choose the required base dimension needed to describe electromagnetic phenomena to be charge, q (the Coulomb in the SI), denoted by $[q] = \mathsf{Q}$, or by a flow of charged particles (an electric current; the ampere in the SI), $[I] = \mathsf{TI} =$ A s in the SI. The various systems of units encountered in electromagnetism are the result of a long history. With the coulomb, rather than the ampere as the realization of this additional dimension (outside of the three mechanical dimensions), some of the basic quantities encountered in electromagnetism: current (I), voltage or EMF (V), capacitance (C), resistance (R) and magnetic field (B) have dimensions: $[I] = \mathsf{QT^{-1}}$, $[V] = \mathsf{ML^2Q^{-1}T^{-2}}$, $[C] = \mathsf{Q^2T^2M^{-1}L^{-2}}$, $[R] = \mathsf{ML^2Q^{-2}T^{-1}}$ and $[B] = \mathsf{MQ^{-1}T^{-1}}$, respectively. If we consider the flow of charged particles to be the basis for the dimension of electric current (see table 1.3 of SI definitions), then for these same quantities, we would write: $[I] = \mathsf{TI}$, $[V] = \mathsf{ML^2\ T^{-3}\ I^{-1}}$, $[C] = \mathsf{M^{-1}\ L^{-2}\ T^4\ I^2}$, $[R] = \mathsf{ML^2\ T^{-3}\ I^{-2}}$ and $[B] = \mathsf{MT^{-2}\ I^{-1}}$, respectively.

In the SI-description of electric and magnetic fields, we can use Gauss' law and Ampere's law; namely, $\nabla . \boldsymbol{E} = \rho/\varepsilon_0$ and $\nabla \wedge \boldsymbol{B} = \mu_0 \boldsymbol{J}$, or the electric and magnetic force laws of Coulomb and Biot–Savart, respectively, to describe two new fundamental constants governing electricity (ε_0) and magnetism (μ_0). The extension of these results by Maxwell led to another great unification of physics; bringing together electricity, magnetism and optics, with their fundamental constants being related by the speed of light, c: $c = 1/\sqrt{\varepsilon_0\mu_0}$.

Thermal physics, covering both thermodynamics and statistical mechanics, introduces another base dimension, and a new fundamental constant of Nature. The concept of absolute temperature, T_K, or T, as described by the ideal gas law, $PV \propto T$, requires a new dimension, which is conventionally written as $[T_K] = \theta$. This is the Boltzmann constant k_B, appearing in the ideal gas law; $[k_B] = \frac{\text{Energy}}{\text{Temperature}} = \mathsf{ML^2T^{-2}}\ \theta^{-1} = \mathrm{J\ K^{-1}}$. The unification here is the connection between the microstates of a physical system, and its bulk macroscopic properties.

Leading on directly from a consideration of how we couple the macroscopic world to the atomic world, is a consideration of quantum mechanics. Although quantum mechanics and relativity are newer ways of examining Nature; when compared with classical mechanics they do not require any additional dimensional quantities. A discussion of modern physics does, however, introduce two fundamental constants: the Planck constant, h, and the speed of light, c. The latter constant is simply a velocity, and so its dimensions are $\mathsf{LT^{-1}}$. The Planck constant is, however, more subtle. At the end of the 19th Century, Max Planck attempted to measure the energy content of the radiation emitted by a black-body (see chapter 10 for details). In fact, what he was doing was questioning the nature of measurement itself. If you wish to measure the temperature of a body; how small a perturbation of that body can you make and still obtain the measurement you desire, without having changed the body you are studying to such an extent that your measurement becomes meaningless? What is the size and nature of this tiny amount of energy that you can remove from a body without changing the body irredeemably?

This investigation of the measurement of tiny fluctuations of temperature and energy, both of which are related through the Boltzmann constant, led to the development of new technologies for making such measurements. This in turn led to the desire to make increasingly precise measurements, which triggered the development of even more precise techniques of measurement, and so on. The scientists' questions were driving the development of technology, by always seeking to push the measurement horizon. This research on temperature showed us in detail how energy flows through matter, and led Max Planck to the concept of the quantum, or the smallest possible quantity of energy, and then to quantum mechanics; arguably the most successful theory ever devised in the physical sciences. Planck showed that the energy content (E) of radiation was proportional to the frequency of that radiation (ν); that is $E \propto \nu$. The constant that turns this relationship of proportionality into a relationship of equality is today known as the Planck constant; that is, $E = h\nu$. And the dimensions of h would be the same as E/ν = energy/frequency = $\mathsf{ML^2T^{-2}/T^{-1}}$; the most precise value we have today (it was fixed in 2019 to redefine the kilogram in the Quantum-SI) is $6.626\ 070\ 15 \times 10^{-34}$ J s.

IOP Publishing

Dimensional Analysis
The great principle of similitude
Jeffrey H Williams

Chapter 1

The origin of units

'I have no satisfaction in formulas unless I feel their arithmetical magnitude.'
William Thomson (created Baron Kelvin, 1892).
From lecture at Johns Hopkins University, Baltimore (1884)

As pointed out in the *Introduction*, science may be thought of as a language adapted to examine and describe Nature. The first attempt at producing a true technical language, based on philosophic principles that might be used to communicate science was due to the polymath, the Reverend John Wilkins (1614–1672), one of the pre-eminent scientific innovators of that period—Wilkins assisted in the founding of the Royal Society of London. In 1668, Wilkins published his *Essay towards a Real Character and a Philosophical Language*, where he attempted to create a universal language to replace Latin as an unambiguous means of communication for international scholarship. The *Essay* also proposed ideas on weights and measures which are similar to those found later in the Metric System of April, 1795. In particular, Wilkins suggested that a universal system of weights and measures could be constructed by using the decimal metric system, based upon a measurement of length.

John Wilkins spoke of a single, universal measure upon which all other measures could be based, and from which all other measures could be derived by mathematics. Wilkins' *Essay* was translated into Italian in 1675 by Tito Livio Burattini (1617–1681), who translated Wilkins' phrase 'universal measure' as *metro cattolico*; thereby introducing the familiar modern word, metre. By the end of the 17th Century, ideas of a rational, universal language were very much part of the European *Zeitgeist*. In 1666, the German polymath Gottfried Wilhelm Leibnitz (1646–1716) published his *Dissertatio de arte combinatoria* in which he claimed that a true philosophical language would be able to analyse all possible concepts into their simplest elements; into what Leibnitz termed 'the alphabet of thought'. In such a philosophical or, as we would say today, a scientific language, a proper symbol should indicate the nature of the animal, phenomenon or whatever it was naming. In other words, it was a language, which could define that thing, or that phenomenon by means of that thing's appearance, or that phenomenon's intrinsic properties. Leibnitz was writing in a century which had attempted the construction of many universal philosophical languages, and so presupposed that a complete enumeration of human knowledge could be achieved. The question that arises immediately is how a relatively small number of base units, or dimensions, could be manipulated or combined to produce a true universal scientific language capable of describing all Nature?

At the time of the French Revolution, *savants* who were familiar with the ideas of Wilkins assumed that the new fundamental unit of length, the metre, could be used to define all the scientific and technological concepts required by their society. This metre was defined from the dimensions of the Earth, as one ten millionth of a quadrant of the Earth's circumference; and is one of the seven base quantities of the International System of Units (the SI), see tables 1.1 and 1.3.

Having defined the basic quantity, and unit of length, l, distances could now be expressed in multiples of this unit. Then to define an area of land, a two-dimensional quantity, you simply multiplied two distances, each expressed in metres. Similarly, three spatial dimensions define a volume as so many cubic metres. To go further, we may, for example, assume that the density (the mass of a known volume of a substance) of pure water is well-defined as, for example, one gram for each cubic centimetre; one can then define a base unit of mass as the weight of a precisely-known volume of pure water. The kilogram was originally defined as the mass of 1000 cubic centimetres, a litre, or a cubic box where the sides are each of 10 centimetres, of pure water.

Table 1.1. Base, or fundamental quantities and their dimensions; together with the appropriate base units in the SI.

Base quantity	Symbol	Description	SI base unit	Dimension
Length	l	The one-dimensional extent of an object	metre (m)	L
Mass	m	A measure of the resistance to acceleration	kilogram (kg)	M
Time	t	The duration of an event	second (s)	T
Electric current	I	Rate of flow of charge per unit time	ampere (A)	I
Temperature	T	Average kinetic energy, per degree of freedom of a system	kelvin (K)	θ
Amount of substance	n	The quantity proportional to the number of particles in a sample, with the Avogadro constant as the proportionality constant	mole (mol)	N
Luminous intensity	I_v	Wavelength-weighted power of emitted light per unit solid angle	candela (cd)	J

But what happens when we wish to consider the combination of length with other quantities which are essential in even some of the simplest concepts of technology; for example, how does one introduce time into a system of mechanical quantities? The speed, or velocity of a comet flying through space, or of a man ploughing a field is defined in terms of distance and time, yet how do we combine these two different base units? One might think of these two quantities as being as different as apples and oranges, so how can they be divided or multiplied together; they certainly cannot be added or subtracted? It is the mathematical definition of a unit that allows us to manipulate distance and time, and generate new ideas such as the concept of speed, and of acceleration. First, consider what we mean by a unit. Any value of a physical quantity, Q, may be expressed as the product of a unit $[Q]$ and a purely numerical factor, or value (a simple number). Written algebraically, we have $Q =$ (a number) $\times$ $[Q]$, where $[Q]$ is the unit; for example, there are a certain number of metres or seconds, $Q_{\text{length}} = 10$ metres or $Q_{\text{time}} = 10$ seconds.

This convention of expressing a quantity as a unit, and a numerical factor is used throughout science, and is referred to as quantity calculus, and is the subject of this volume. When units are being manipulated, one may only add like terms, as with apples and oranges, but all units may be manipulated algebraically. When a unit is divided by itself (that is, metres/metres), the division yields a dimensionless number, which is one (1, or unity) and so intrinsically without dimension or a unit. When two different units are multiplied or divided, the result is always a new unit, referred to by the combination of the individual units. For instance, in the SI, the unit of velocity is metres per second; this unit is neither length, nor is it time, but length divided by time.

Length and time are base quantities, and base dimensions; that is, they cannot be decomposed into simpler components, but the new unit of velocity is said to be a derived unit containing more than one base dimension, and may be deconstructed into base units, and into base dimensions (length, L, and time, T). In the same way, the density of water we encountered above in the definition of the base unit of mass is also a derived unit. Density is defined as the mass of a known volume, or mass per unit volume; and is composed of two base quantities, the base unit of mass (kilogram, of base dimension, M) and the base unit of length (metre, of base dimension, L), which as we are dealing with a volume is cubed (kilogram/metre^3, or kg m^{-3}).

As science and technology advanced in the 19th Century, the new profession of scientists understood how the various manifestations of, for example, heat and work were all related to the concept of energy, and how energy related to the established base quantities of length, mass and time. In fact, today we have seven base quantities, which may be combined to explain every known scientific phenomenon, and which would be used to comprehend scientific discoveries that have yet to be made. That is, it is through these seven base quantities that the true universal language, the language of science is formulated. Each of the seven base quantities used in the SI is regarded as having its own dimension. Table 1.1 contains the seven base quantities of the SI, together with their descriptions, and their respective dimensions.

All the phenomena known to science are derived from these seven base quantities, contained in table 1.1 (see the selection given in table 1.2), using the well-established equations or laws of physics, and are termed derived quantities. The dimensions of the derived quantities are written as products of powers of the dimensions of the base quantities using the equations that relate the derived quantities to the base quantities.

In the following chapters, we will see that quantity calculus and dimensional analysis are powerful tools in understanding the properties of physical quantities independent of the units used to measure them, and is the manner in which we use the language of science to define and measure Nature. Every physical quantity is some combination of the quantities in table 1.1. Dimensional symbols and exponents are manipulated using the rules of algebra; for example, the dimension of area is written as L^2, the dimension of velocity as LT^{-1}, the dimension of acceleration (the rate of change of velocity with respect to time) is written as LT^{-2}, and the dimension of density as ML^{-3}. Table 1.2 is a selection of physical quantities, from a number of fields of physical science together with the appropriate SI derived unit and their dimensions.

We will see that dimensional analysis is routinely used to check the plausibility of newly-derived equations, the design of experiments, and the results of calculations in engineering and science before resources and effort are expended on detailed measurements. In this way, reasonable hypotheses about complex physical situations are examined theoretically, to see if they merit subsequent testing by experiment. And, it is also the means by which one seeks to determine equivalent values for a quantity in another system of units. For example, how you convert from the value of a quantity in metric units to the equivalent quantity in British customary units: metres/second to miles/hour, or joules (the SI derived unit of energy, symbol J, where J is equivalent to kg $\mathrm{m}^2\ \mathrm{s}^{-2}$) to British Thermal Units or BTU (a customary unit of energy equal to about 1055 joules. A BTU is approximately the amount of energy needed to heat one pound (0.454 kg) of water, which is exactly one tenth of a UK gallon or about 0.1198 US gallons, from 39 °F to 40 °F or 3.8 °C to 4.4 °C). Thus, dimensional analysis is the means of translating between the various dialects of the single universal language of science.

1.1 The *système internationale des unités* (the *SI*)

As mentioned above, the derived units of the SI are constructed algebraically from the base units (composed of base dimensions) of the SI, using the law of physics. However, sometimes a choice arises between different equations governing the phenomenon of interest, and the manner in which a derived unit may be constructed. An important example occurs in the manner of defining the quantities describing electricity and magnetism.

The theory of electricity and magnetism (which became electromagnetism in the late-19th Century) was developed during the early-19th Century as discoveries were made by Hans Christian Ørsted, André-Marie Ampère and Michael Faraday. The possibility of defining magnetic phenomena in terms of the mechanical dimensions

Table 1.2. A selection of physical quantities, together with their SI derived unit and their dimensions. (Note: It is the dimension that is used in dimensional analysis, not the SI unit.)

Physical quantity	Symbol	Description	SI derived unit	Dimension
Angular momentum	L	Measure of the extent and direction an object rotates about a reference point	kg m^2 s^{-1}	M L^2 T^{-1}
Dynamic viscosity	v	Measure for the resistance of an incompressible fluid to stress	Pa s	M L^{-1} T^{-1}
Electric field strength	E	Strength of the electric field	V m^{-1}	M L T^{-3} I^{-1}
Electric potential	φ	Energy required to move a unit charge through an electric field from a reference point	volt (V = J C^{-1})	M L^2 T^{-3} I^{-1}
Electrical resistance	R	Electric potential per unit electric current	ohm (Ω = V A^{-1})	M L^2 T^{-3} I^{-2}
Energy	E	Energy	J	M L^2 T^{-2}
Entropy	S	Logarithmic measure of the number of available states of a system	J K^{-1}	M L^2 T^{-2} θ^{-1}
Force	F	Transfer of momentum per unit time	newton (N = kg m s^{-2})	M L T^{-2}
Heat capacity	C_p	Energy per unit temperature change	J K^{-1}	M L^2 T^{-2} θ^{-1}
Illuminance	E_v	Luminous flux per unit surface area	lux (lx = cd Sr m^{-2})	L^{-2} J
Irradiance	E	Electromagnetic radiation power per unit surface area	W m^{-2}	M T^{-3}
Intensity	I	Power per unit cross-sectional area	W m^{-2}	M T^{-3}
Magnetic field strength	H	Strength of a magnetic field	A m^{-1}	L^{-1} I
Magnetic flux	Φ	Measure of magnetism, taking account of the strength and the extent of a magnetic field	weber (Wb)	M L^2 T^{-2} I^{-1}
Molar concentration	C	Amount of substance per unit volume	mol m^{-3}	N L^{-3}

(*Continued*)

Table 1.2. (*Continued*)

Physical quantity	Symbol	Description	SI derived unit	Dimension
Permeability	μ_s	Measure for how the magnetization of material is affected by the application of an external magnetic field	H m^{-1}	M L T^{-2} I^{-2}
Permittivity	ε_s	Measure for how the polarization of a material is affected by the application of an external electric field	F m^{-1}	M^{-1} L^{-3} T^4 I^2
Power	P	Rate of transfer of energy per unit time	watt (W)	M L^2 T^{-3}
Pressure	p	Force per unit area	pascal (Pa = N m^{-2})	M L^{-1} T^{-2}
Radiant intensity	I	Power of emitted electromagnetic radiation per unit solid angle	W Sr^{-1} ($m^4 m^{-2}$ kg s^{-3} = m^2 kg s^{-2})	M L^2 T^{-3}
Solid angle	Ω	Ratio of area on a sphere to its radius squared	steradian (Sr) (m^2 m^{-2})	1
Surface tension	γ	Energy change per unit change in surface area	N m^{-1} or J m^{-2}	M T^{-2}
Thermal conductivity	λ	Measure for the ease with which a material conducts heat	W (m K)$^{-1}$	M L T^{-3} θ^{-1}
Wavelength	λ	Perpendicular distance between repeating units of a wave	m	L
Weight	w	Gravitational force on an object	newton (N = kg m s^{-2})	M L T^{-2}

of length, mass and time; that is, creating a coherent universal system of units was first suggested in 1833 by the German mathematician Carl Frederick Gauss (1777–1855). His analysis was extended to cover electrical phenomena by Wilhelm Weber (1804–1891), who in 1851 proposed a method by which a complete set of units could be constructed which incorporated electromagnetism into the existing mechanical Metric System.

Scientists have spent more than a century disagreeing about the units for electromagnetism. But even if every scientist were suddenly to adopt SI units for electromagnetism, the need for familiarity with the older c.g.s. (centimetre-gram-second) system of units would still exist, so as to facilitate reading of the vast published literature in this area of science since the early-19th Century.

The problem begins with deciding what exactly constitutes an electric or a magnetic field. In 1861, a committee of the British Association for the Advancement of Science (BAAS) that included William Thomson (later Lord Kelvin), James Clerk Maxwell and James Prescott Joule, undertook a comprehensive study of electrical measurements. This committee introduced the concept of a comprehensive system of units for electromagnetism. It was discovered that only four simple equations were required to define and couple the units of electrical charge Q, electric current I, voltage or potential difference V, and electrical resistance R. These simple equations are:

- **either** Coulomb's force law (named after Charles-Augustin de Coulomb,) for charges, which tells us that like charges repel each other and unlike charges attract each other, **or** Ampère's force law (named after André-Marie Ampère) for currents (that a repulsive force will develop between two neighbouring parallel wires through which electrical currents are passing in the same direction). Ampère's law was the basis for the old definition for the SI base unit of electric current (see table 1.3).
- The relationship between electric charge and electric current that tells us that the charge (Q) passed through a conductor is equal to the electrical current (I) passed multiplied by the time for which the current flows (T); that is, $Q = IT$ (this is the basis of the new SI definition of the ampere, see table 1.3).
- The well-known law due to Georg Simon Ohm (1789–1854) that an electric current (I) passing through a wire with a certain electrical resistance (R) will develop a voltage (V), such that $V = I\,R$ (Ohm's law).
- The equation for electrical work; that the work (W) done or energy expended when an electrical current (I) passes through a wire of resistance (R) for a time T is given by $W = I^2RT$.

A fundamental principle of the new system of units desired by those 19th Century scientists, seeking to incorporate electromagnetism into the existing systems of mechanical units, was that the new system of units should be coherent. That is, the system must be based upon certain base units for fundamental dimensions such as length, mass and time. Then derived units could be constructed for a vast number of other scientific quantities as products or quotients by means of quantity calculus, but without requiring additional numerical factors. The metre, gram and second were originally selected as base units by the BAAS committee in 1861. Consequently, this was not a system of units based on the Metric System, which has the kilogram not the gram as the unit of mass, and so there is a complicating numerical factor of 10^3.

Because one could use either the force law due to Coulomb, or the force law due to Ampère to define electromagnetic quantities, two parallel systems of units arose, the electrostatic and electromagnetic subsystems of the c.g.s. system of units, depending on whether the law of force for electric charges (Coulomb's Law) or the law for electric currents (Ampère's Law) was used as the origin of these forces.

Table 1.3. Definitions of the base units of the SI before 20 May, 2019 (the old-SI), and subsequent to that date (the Quantum-SI). A full explanation of the changes made in 2019 may be found in [1].

Base quantity	Definition, pre-May 2019	Present definition
		The International System of Units, the SI, is the system of units in which:
Metre The base unit of length is the metre (m)	The metre is the length of the path travelled by light in vacuum during a time interval of 1/299 792 458 of a second	The speed of light in vacuum c is 299 792 458 ms^{-1}
Kilogram The base unit of mass is the kilogram (kg)	The kilogram is equal to the mass of the International Prototype of the Kilogram	The Planck constant h is 6.626 070 15 × 10^{-34} J s
Second The base unit of time is the second (s)	The second is the duration of 9 192 631 770 periods of the radiation corresponding to the transition between the two hyperfine levels of the ground state of the Caesium-133 atom. (This definition refers to a Caesium atom at rest at a temperature of 0 K)	The unperturbed ground state hyperfine transition frequency of the Caesium-133 atom $\Delta\nu_{Cs}$ is 9 192 631 770 Hz
Ampere The base unit of electric current is the ampere (A)	The ampere is that constant current which, if maintained in two straight parallel conductors of infinite length, of negligible circular cross-section, and placed 1 metre apart in vacuum, would produce between these conductors a force equal to 2 × 10^{-7} newton per metre of length	The elementary charge e is 1.602 176 634 × 10^{-19} C
Kelvin The base unit of thermodynamic temperature is the kelvin (K)	The kelvin, the unit of thermodynamic temperature, is the fraction 1/273.16 of the thermodynamic temperature of the triple-point of water. (This definition refers to water of the isotopic composition: 0.000 155 76 mole of ^{2}H per mole of ^{1}H, 0.000 379 9 mole of ^{17}O per mole of ^{16}O, and 0.002 005 2 mole of ^{18}O per mole of ^{16}O)	The Boltzmann constant k_B is 1.380 649 × 10^{-23} J K^{-1}

Candela The base unit of light intensity is the candela (cd)	The candela is the luminous intensity, in a given direction, of a source that emits monochromatic radiation of frequency 540×10^{12} hertz and that has a radiant intensity in that direction of 1/683 watt per steradian.	The luminous efficacy of monochromatic radiation of frequency 540×10^{12} Hz, K_{cd}, is 683 $\mathrm{lm\ W^{-1}}$
Mole The base unit of amount of substance is the mole (mol)	The mole is the amount of substance of a system which contains as many elementary entities as there are atoms in 0.012 kilogram of Carbon-12.	The Avogadro constant N_A is $6.022\,140\,76 \times 10^{23}\ \mathrm{mol^{-1}}$

In 1888, the German physicist Heinrich Hertz (1857–1894) verified Maxwell's predictions of the nature of electromagnetic radiation and light. His experiments greatly extended the range of frequencies over which Maxwell's equations were shown to be valid. In addition, Hertz combined the electrostatic and electromagnetic subsystems of the c.g.s. system of units into a single system where the various terms were related by the speed of light *c*; he called this new system the Gaussian system of units in honour of Karl Friedrich Gauss.

The recommendations of the BAAS of 1873 were adopted by the first International Electrical Congress in Paris in 1881. Five practical electrical units were defined: the ohm (Ω), named for George Ohm, the farad (F), named for Michael Faraday, the volt (V), named for Alessandro Volta, the ampere (A), named for André-Marie Ampère, and the coulomb (C), named for Charles-Augustin de Coulomb. In 1889, the second International Electrical Congress added to this list of units: the joule (J) to name a quantity of energy, in honour of James Prescott Joule, the watt (W) to name a quantity of work, in honour of James Watt, and a unit of electromagnetic inductance, later given the name henry (H) in honour of the American physicist Joseph Henry (1797–1878).

In 1901, an Italian electrical engineer, Giovanni Giorgi (1871–1950), demonstrated that the BAAS electrical units and the m.k.s. (**m**etre-**k**ilogram-**s**econd) mechanical units (derived from the Metric System), introduced by the 1st CGPM (*Conférence générale des poids et mesures*) of 1889, could be incorporated into a single coherent system of units by selecting the metre, kilogram and second as the base units for mechanical quantities, and adding a base unit of an electrical nature, and, finally, by giving a physical character to the space through which the electric or magnetic field propagated. This latter point required the assignment of physical dimensions to the permeability of free space (this is a property of the vacuum) through which electric and magnetic fields propagate (see the definitions of the base quantities of the SI in table 1.3).

With the need for further international cooperation on units of electromagnetism, the 6th CGPM of 1921 amended the Metre Convention of 1875 so that units of electromagnetism would be regulated by the institutions created by the Metre Convention. By the 8th CGPM in 1933, there was an international desire to assist the electricity industry by creating an absolute set of electrical units based on Giorgi's proposals, with the practical electrical units incorporated into a comprehensive m.k.s. system of units.

In September 1935, the world of measurement science accepted that the ampere should be chosen as the base unit for electricity, and that it be defined in terms of the force per unit length between two long parallel wires (see table 1.3). By 1946, international organizations with an interest in the international coordination of the physical sciences accepted Giorgi's proposal for a four-dimensional system of units based on the **m**etre-**k**ilogram-**s**econd-**a**mpere (m.k.s.a. system of units), to take effect on 1 January 1948.

Following the mechanism for international metrological collaboration created under the Metre Convention, the CGPM instructed the *Comite internationale des poids et mesures* (CIPM) to investigate the creation of a complete set of units of measurement. In addition, the CIPM was instructed to investigate the opinion prevailing in the Member States of the Metre Convention on the creation of a modern system of units based on the Metric System. These moves meant that the m.k.s.a. system quickly evolved into the *Système international d'unités*, or SI (International System of Units) of 1960.

In summary, the various stages of the evolution of the systems of units, which we today call the SI, and which derived from the Metric System of 1795 are as follows:

- 1668: Bishop John Wilkins' proposal for a universal measure.
- 1795: The Metric System created in Revolutionary France.
- The c.g.s. metric system of about 1872, which then split into an electrostatic-c.g.s. system and a magnetic-c.g.s. system of units, depending upon the choice of Ampère's Law or Coulomb's Law to define force.
- 1875: The Metric System was proposed to all nations through the Metre Convention.
- 1901: Giorgi's m.k.s.a. proposal for a coherent system of metric units.
- 1948: Giorgi's m.k.s.a. system was formally adopted, and the other metric systems of units were abandoned.
- 1960: Creation of the SI.
- 2019: Redefinition of the base units of the SI, and the creation of the Quantum-SI [1]

As with all living languages, the language of science is forever evolving; and just as with the evolution of a natural language, the direction the evolution takes does not always please all those who use the language. However, this evolution is a sign of its health and utility. The most recent change to the SI occurred in 2019 [1] with the creation of the Quantum-SI. This change came about from the need to replace the artefact-based definition of the kilogram. To create the Quantum-SI, the definitions of four of the base quantities changed (table 1.3 gives the old and the new definitions); and the new definitions are based on fixed values for some fundamental constants of Nature. In table 1.3, notice the different levels of precision of the constants of Nature that define the base quantities in the new-SI (the frequency of an atomic Caesium clock is known to a part in 10^{10}, and the speed of light is quoted to nine significant figures), but the constant that defines the candela is singularly lacking in precision—the luminous efficacy is known only to a part in a thousand. But the search for greater precision is at the heart of the evolution of physics.

Further reading

A fascinating, and very readable account of the philosophical languages proposed by John Wilkins and Gottfried Leibniz, and from which modern ideas of rational systems of weights and measures evolved is Eco U 1997 *The Search for the Perfect Language* London: Fontana Press

Reference

[1] Williams J H 2020 *Defining and Measuring Nature: The Make of all Things* 2nd edn (Bristol: IOP Publishing)

Chapter 2

A brief history of dimensional analysis: a holistic approach to physics

> *'If two particles are carried with uniform motion, but each with a different speed, the distances covered by them during unequal intervals of time bear to each other the compound ratio of the speeds and time intervals.'*
>
> Galileo Galilei (1564–1642)
> (Proposition IV, *Third Day*)

A contemporary scientist would represent the above statement by writing $d = vt$; meaning distance travelled (dimension, L) is the product of velocity, v (dimension, LT^{-1}) and time, t (dimension, T). In the works of the early-physicists, Galileo, Christiaan Huygens (1629–1695) and Isaac Newton (1643–1727), one never finds a numerical value for a velocity, an acceleration or a force. Instead, results are stated in terms of ratios (as in the above quotation). Any calculation that one could perform in the modern manner, using equations that relate dimensioned quantities would have been performed in Galileo's day using ratios. However, the modern method has significant advantages: it readily permits conversion between different systems of units of measurement. And it is much more concise; enabling dimensional analysis, where one may infer the form of a physical equation.

Below, we will briefly explore the development of the concept of units and dimensions in physics. There was a gradual transition from Galileo's ratios to equations that relate dimensioned quantities. The first scientist to have used the modern understanding of dimensioned quantities was the French mathematician, Joseph Fourier (1768–1830), who stated these principles in the later editions of his *The Analytic Theory of Heat* [1].

doi:10.1088/978-0-7503-3655-0ch2

2.1 Homogeneity of units

The principle of homogeneity states that if $A = B$ is physically meaningful, then A must have the same units as B. Similarly, if $A + B$ appears in an equation, then A must have the same units as B. For example, it does not make sense to add a length to an area; and section 3.3 outlines the disastrous consequence of mixing systems of units. An early statement of this principle of homogeneity can be found in Euclid's *Elements*, where Euclid states that '*only things of the same kind can be compared to each other*,' where angles, lengths, areas and volumes are different kinds.

Greek mathematics was based on geometry, and arithmetic expressions were considered to be sums and products of lengths and areas; that is, everything had a geometrical interpretation. René Descartes (1596–1650) broke this Greek tradition when he developed analytic geometry. He considered curves in the plane described by equations such as $y = x^3 - 3x^2$, and he assumed that x and y were lengths. Descartes justified these equations by stating that the higher-order terms (x^2 and x^3) are implicitly multiplied by some power of the unit length u, so that the equation actually reads $y = (x^3/u^2) - (3x^2/u)$, and all terms have the dimension of length. This means that analytic geometry requires a reference length, the unit length; whereas in Euclidean geometry there is no preferred length.

2.2 Geometry of motion

Galileo Galilei's *Two New Sciences* [2] is mostly concerned with kinematics: describing trajectories and times of flight of objects moving without air resistance. Galileo speaks of the 'speed' of particles, but his results are stated in terms of ratios of distances and times.

Galileo did not think it was sensible to take the product of a speed, or velocity and a time, as we do in the modern statement of this proposition: $d = vt$. Galileo never speaks of numerical values of speeds, though he refers to the ratios of speeds of non-accelerated objects, and the average speeds of accelerated objects; he talks about an equivalent object that undergoes uniform, non-accelerated motion. For example, Galileo wrote: '*Concerning motions and their velocities or momenta, whether uniform or naturally accelerated, one cannot speak definitely unless he has established a measure for such velocities and also for time. As for time we have already widely adopted hours, first minutes, and second minutes* [the second is humanity's oldest universal unit of measurement]. *So, for velocities, just as for intervals of time, there is need of a common standard which shall be understood and accepted by everyone, and which shall be the same for all… Let us consider the speed and momentum acquired by a body falling through the height, say, of a spear, as a standard which we may use in the measurements of other speeds and momenta as occasion demands.*'

These deliberations about the need for a widely-accepted unit for speed sound peculiar to us, but we live in a world where time is more efficiently used (and measured) than four centuries ago—although his inference about the need for accepted units of distance, is ahead of its time. Firstly, distance per time has, at least, two possible meanings: the distance travelled by an object uniformly accelerated from rest, and the distance travelled by a non-accelerated object with constant speed. Galileo uses the former meaning in his definition of velocity, possibly because it is a much easier way of reliably producing a given speed. While Galileo seems to have an understanding of the instantaneous velocity of an accelerated object, this concept was not formal and intuitive before the development of calculus. Secondly, the idea of dividing units would have been unknown at the time of Galileo. A modern science student takes it for granted that the objects he/she manipulates in physics are not the numbers of arithmetic, but a unit and a numerical portion (see chapter 1). Thirdly, there were several systems for measuring distance, none of which were standardized, so one may speculate that scientists preferred to write of ratios of lengths rather than lengths (customary to London, Paris or Florence) to make their results more reproducible.

The later parts of Galileo's *Two New Sciences* contain derivations of some results. The modern approach would involve algebraic manipulation of dimensioned quantities, but Galileo (and Newton) had a different approach: they drew geometric figures where different points represented positions, velocities and times. Consider: '*Let the motion take place along the line ab, starting from rest at a, and in this line choose any point c. Let ac represent the time, or the measure of the time, required for the body to fall through space ac, let ac also represent the velocity at c acquired by a fall through the distance ac.*' (Proposition III, *Fourth Day*). The length *ac* is being used as a length (L), a velocity (LT^{-1}), and a time (T). Galileo is effectively choosing units for length and time, so that $d = \frac{1}{2}\, gt^2 = gt = t$; that is, units where the gravitational acceleration, $g = 1$ and $t = \sqrt{2}$. Then he derives, for example, that a certain point *s* (between *c* and *b*) is the velocity of the object when its position is *b*. In the next section, Galileo uses these geometric techniques to derive various properties of parabolic trajectories. Problem 1, Proposition IV is '*To determine the momentum of a projectile at each point in its given parabolic path.*' So, Galileo did have the concept of an instantaneous velocity.

In his *Principia Mathematica* [3] (of 1687), Isaac Newton speaks of forces and other dimensioned quantities. Newton treats forces as Galileo treated velocities: he acknowledges that force has a physical meaning, and he talks about ratios of forces; but he never talks about the numerical value of a force, and he never writes an equation where force appears alone on one side. The modern statement of Newton's second law is the well-known, $F = ma$, but Newton wrote '*A change in motion is proportional to the motive force impressed and takes place along the straight line in which that force is impressed.*' In the section of the *Principia* on the contribution of the Sun and moon to tides, Newton writes, '*Wherefore since the force of the Sun is to the force of gravity as* 1 *to* 12 868 200, *the Moon's force will be to the force of gravity as* 1 *to* 2981 400' (Proposition XXXVII, Problem XVIII, Book 3).

In the 1790s, French *savants* debated about international standards for measurements, and in April 1795 the Metric System became (for a few years) the official system of measurement in France. The second was a conventional, culturally familiar unit for time dating from the period of the Sumerians, but the metre and kilogram were novel. These units were designed so that any laboratory could perform precision measurements to reproduce the standards of length and mass, and related derived units. Such measurement systems were standardized by the fabrication and dissemination of standard prototypes or artefacts.

The original definitions of the metric units were as follows [4]:

- metre: became 1/10 000 000 the distance between the North Pole and the Equator;
- second: 1/86 400 of the mean solar day;
- kilogram: the mass of a litre of pure water; that is, a cube with side 1/10 metre; and
- the Celsius temperature scale was based on the properties of water, where 0 degrees is the freezing point and 100 degrees is the boiling point, at atmospheric pressure.

While it took a long time for these units to be officially adopted by most governments (the international Metre Convention was signed in May 1875), these units were quickly adopted by scientists and engineers. For example, an 1804 article in the English journal *Philosophical Magazine* describes a method of measuring the velocity of a cannonball shot out of a cannon in France [5]. The article gives an equation for the speed in terms of several measured quantities, and lists the results as so many metres per second[1].

This standardization of quantities and units, and the introduction of quantity calculus enabled scientists (particularly, physicists) to write equations involving dimensioned quantities, rather than the ratios employed by Galileo and Newton. This was likely the catalyst that led to the 19th Century development of physics as a distinct subject. The researches of Leonhard Euler (1707–1783), such as his work on ship stability originally published in 1749, is full of equations; some of which contain dimensioned constants such as the acceleration due to gravity, g. So, physics was written as equations before the legal creation of the Metric System. But the existence of the Metric System, and the epistemological earthquake that was the French Revolution certainly made this format more convenient; since one did not have to worry that one was measuring distance in Paris feet while using the value of g measured in English feet per (universal) second squared.

2.3 Derived units

By using the laws of physics as a grammar, and the base units as expressions or words, we may construct a language that allows us to make predictions about phenomena that have not yet been identified, but should be observable. A scientist could ask questions such as: What happens if a force tries to twist a body instead of pushing (repelling) or pulling (attracting) it? What happens if a force acts upon an area, not simply along a line? In the first case, one defines torque, which is a force that tries to rotate a body. In the second case, a force acting over an area would be newtons per square metre; that is, $\text{kg m s}^{-2}/\text{m}^2$ or $\text{kg m}^{-1}\text{s}^{-2}$, which is the definition of pressure. Pressure is nothing more than the force exerted by something (a gas or a fluid) upon a well-defined area. The British customary unit for pressure is pounds per square inch, which gives a clear indication of pressure as a force upon an area. As for

[1] It is of note that in 1804, England and France were at war, and so this publication in the English media of the detailed characteristics of a new French cannon, which was likely a state secret in France, probably involved espionage.

the reasonableness of phenomena that have not yet been observed; one first has to consider the magnitude of the units, and a dimensional analysis to see if such a new phenomenon would be observable. Consider the pressure exerted by light—radiation pressure. Could it exist? Is it measurable? The answer is in the affirmative, and it was discovered that the radiation of the Sun exerts a pressure of less than a billionth of an atmosphere on the Earth's surface. But it was an examination of the existing language of science, which suggested, and allowed individuals to look for this new phenomenon.

Dimensional analysis was known to Isaac Newton who referred to it as the Great Principle of Similitude. The 19th Century French mathematician and Egyptologist Joseph Fourier (1768–1830) made important contributions to dimensional analysis based on the idea that physical laws like Newton's law, $F = m\ a$, should be independent of the systems of units employed to measure the physical variables. That is, the laws of Nature and fundamental equations should be equally valid in the Metric System of units, as in a non-metric system of units. When converting between these two systems of units, we need only be aware of the mathematical factors needed to convert between the base units to convert the entire quantity from one system to another. Thus, one should take care never to mix systems of units, as the consequences could be disastrous (see section 3.3). But there is nothing stopping one defining force in ancient Egyptian units; distance would be in terms of the royal cubit (about 0.525 metre), mass would be in *deben* (about 0.015 kilogram) and time would have been in *unut* (the hour, which is identical to our modern hour). Fourier's ideas show how each of these base units would need to be converted to SI base units to convert the ancient Egyptian unit of force (a *pharaoh*?) to the newton.

2.4 Fourier and the nature of physical quantities

Joseph Fourier was the first to recognize that physical quantities have units associated with them, and that this character determines how these quantities rescale under changes in the system of units employed to quantify them. He presented his ideas in a chapter, *General Remarks* in his major work, *The Analytic Theory of Heat* (Section IX) [1]. Where he states that physical quantities have dimensions: '*In order to measure these quantities and express them numerically, they must be compared with different kinds of units, five in number, namely, the unit of length, the unit of time, that of temperature, that of weight, and finally the unit which serves to measure quantities of heat.*' (Today, we say that Fourier's system of five base quantities (dimensions) are, in fact, only four—see table 2.1.)

Joseph Fourier began by exploring the variables involved in the study of heat diffusion (this led to Fourier's law of heat conduction): conduction internally within a solid body (defined by a coefficient, K) and externally through its surface, or boundary into the environment (defined by a coefficient, h); and specific heat (defined by a coefficient, C). Fourier used these quantities to define the quantity of heat at a point in, or a section of a body. He took the flow of heat to be uniform, and temperature change linear with respect to distance. He drew upon Newton's law of cooling: that the flow of heat through a domain was proportional to the temperature

Table 2.1. Fourier's dimensional analysis of some properties of solids.

Quantity or constant of interest	Dimension of length, L	Dimension of duration, T	Dimension of temperature, θ	Fourier's dimensionality (modern dimensionality)+
Distance	1	0	0	L (L)
Time	0	1	0	T (T)
Temperature	0	0	1	θ (θ)
Thermal conductance (*K*)	−1	−1	−1	$L^{-1}T^{-1}\theta^{-1}$($MLT^{-3}\theta^{-1}$ = $ML^2T^{-2}.L^{-1}T^{-1}\theta^{-1}$)
Thermal transmittance (*h*)	−2	−1	−1	$L^{-2}T^{-1}\theta^{-1}$($MT^{-3}\theta^{-1}$ = $ML^2T^{-2}.L^{-2}T^{-1}\theta^{-1}$)
Specific heat capacity (*C*)	−3	0	1	$L^{-3}\theta^{-1}$($ML^{-1}T^{-2}\theta^{-1}$)++ = $ML^2T^{-2}.\ L^{-3}\theta^{-1}$

+ The difference between the dimensionality established by Fourier, and how we would represent that dimensionality today comes from how we define energy as a product of three base dimensions, ML^2T^{-2}.
++ Volumetric heat capacity, which is J m^{-3} K^{-1}. In each case, to derive the modern dimension from Fourier's dimension, we require the dimension of energy, ML^2T^{-2}.

difference across it. But he became aware of its limitations. In a novel move, Fourier also derived diffusion equations by considering the quantity of heat in a solid bar between two points; thereby obtaining a differential equation, which he differentiated with respect to a length parameter, x.

Joseph Fourier's calculations involve three coefficients (K, h and C), which differ between materials, and have a non-obvious dependence on modern fundamental units: thermal transmittance, thermal conductance and heat capacity, respectively. Fourier's calculations should hold regardless of the units of measurement. But how do these constants change when the measurement units are changed? Fourier realized that these constants must rescale when the investigator changes the system of measurement units. Fourier pointed out that the proper rescaling factor depends on the exponent of the dimension involved in the property of interest. For example, an area and a volume have exponents of dimension 2 and 3 (that is, L^2 and L^3, respectively) with respect to the dimension of length. Fourier's specific heat, C; that is, the heat required to raise the temperature of a volume of the material under consideration by one degree, has exponent of dimension −3 with respect to length, so if the unit of length is multiplied by λ, then $C \rightarrow C\lambda^{-3}$. Fourier provided a table specifying the dimensions of the undetermined constants k, h and C; this original table has been modified as table 2.1, and uses modern names for quantities:

Fourier points out a generalized principle of dimensional homogeneity, which applies to linear equations and differential equations: '*On applying the preceding rule to the different equations and their transformations, it will be found that they are homogenous with respect to each kind of unit, and that the dimension of every angular or exponential quantity is nothing.*' In other words, just as we cannot add a length to an area, we also cannot add quantities A and B, if A has units of per second and B has units of per second squared. To add quantities, the exponent of the ensemble must be zero; that is, the group of variables or quantities must be dimensionless. Fourier's insights on dimensional analysis far surpassed those of anyone who preceded him. Fourier did not, however, realize that the dimensions of the quantities involved could be used to infer the form of the equation relating them—this technique would be used at the end of the 19th Century by Lord Rayleigh [6].

Neither in his analysis, nor in the general discussion did Fourier make any commitment as to the nature of heat; he took it just as a phenomenon, with cold as its opposite. Here he was in accord with the contemporary Swiss physicist Pierre Prevost. However, in 1807 this view of the nature of heat (a huge controversy at the time) disappointed Laplace, who in 1810 re-derived Fourier's diffusion equation construing heat as a central action between molecules, which was known only to decline rapidly with distance from its source. Fourier neither affirmed nor rejected this method. Joseph Fourier's book became a standard source for both students of heat diffusion, and of Fourier series and integrals (which he invented to solve the problems he was encountering), and the general solution of linear differential equations. The renown of Fourier's achievement spread quickly; William Thomson (later Lord Kelvin) when aged 16 years in 1840, took the book away with him as holiday reading.

2.5 Dimensional arguments

By describing physics with systems of units, we can use the technique of dimensional analysis. We can infer what a physical relation will look like just by knowing the base dimensions involved in the phenomenon of interest. The first such dimensional argument is due to the French military engineer François Daviet de Foncenex (1734–1799). In 1761, de Foncenex considered the total force, R, that results when two forces, with equal magnitude, F, and an angle θ between them are applied to the same object. That is, $R = f(F, \theta)$ for some function f. Since R and F have the same units, but θ is dimensionless, the relation must take the form $R = Fg(\theta)$, for some function g. While this may seem obvious, it was a major advance at a time when no one spoke about the dimensions of force. This idea of de Foncenex was cited by Legendre, Poisson and Fourier.

Force is a vector quantity, which means that it has a magnitude and a direction. Graphically, a force is represented as a line segment from its point of application A to a point B, which defines its direction and magnitude. The length of the segment AB represents the magnitude of the force. Vector calculus was developed in the late-1800s and early-1900s. The parallelogram rule used for the addition of forces, however, dates from antiquity and is noted explicitly by Galileo and Newton. Today, when presented by a problem in astrophysics, classical mechanics, or even quantum mechanics, we seek to examine how the forces, involved in the phenomenon under investigation couple together. Galileo and Newton would, of course, have known how to couple forces but they chose not to use the concept of individual

Table 2.2. Dimensions of variables.

Variable	Dimension
f	T^{-1}
m	M
k	MT^{-2}
A	L

forces interacting on an object (they dealt with ratios of forces); it was Daviet de Foncenex who first used the modern approach to such problems—of seeking a sum of the forces in his problem. Finding such a balance of the appropriate forces is also the means by which Niels Bohr gave us our first quantum picture of the atom (see section 12.2), and is the basis of electromagnetism.

The first modern scientist to use dimensional arguments was John William Strutt, 3rd Baron Rayleigh, who used them extensively in his *Theory of Sound* (1877) [6], which contained a section entitled, *Method of Dimensions.* A simple example of dimensional analysis is as follows: We wish to find the frequency, f, of the small oscillations of a mass (m) at the end of a massless spring. The frequency, f, is a function of the mass, m, the spring constant k, and the amplitude of the oscillations, A. We assume the unknown quantity, f, is proportional to some function containing the variables known to be involved in the problem; that is,

$$f \propto (m, k, A) = C\ g(m, k, A), \tag{2.1}$$

where C is an unknown constant of proportionality. The dimensions of these quantities (where M; L; T are mass, length and time) are given in table 2.2.

We see that $g(m, k, A)$ must have dimension T^{-1}. Lord Rayleigh's argument was that the relation, $f = g(m, k, A)$ should hold regardless of the choice of units; thus both sides must rescale in the same way when we change units. To do this, we must find a, b and c such that

$$\mathsf{T}^{-1} = \mathsf{M}^{a}(\mathsf{MT}^{-2})^{b}\mathsf{L}^{c} = \mathsf{M}^{a+b}\mathsf{T}^{-2b}\mathsf{L}^{c}, \tag{2.2}$$

which is a set of linear equations (one for each base dimension) that must be solved, giving the unique solution: $a = -½$, $b = ½$ and $c = 0$. Thus,

$$f = Ck^{1/2}m^{1/2} = C\sqrt{k/m}. \tag{2.3}$$

Dimensional analysis cannot tell us about the constant, but the exact formula is the well-known $2\pi f = \sqrt{k/m}$ (see chapter 6 for further details). This technique of equating exponents of powers of dimensions, to better understand a new observation is often termed the Rayleigh algorithm.

In the 1880s and 1890s, French scientists began to develop a more general and rigorous theory of dimensional analysis. The engineer, Aimé Vaschy (1857–1899) wrote in 1896: '*Any homogenous relationship among p quantities a_1, a_2, …, a_p, the values of which depend on the choice of the units, may be reduced to a relationship* *involving (p-k) parameters that are monomial combinations of a_1, a_2, …, a_p and that are of zero dimensions* ($L^0M^0T^0 = 1$).' Vaschy's theorem was forgotten. However, it was rediscovered, probably independently, by Dimitri Pavlovitch Riabouchinsky (1882–1962) in Russia in 1911, and by the American Edgar Buckingham (1867–1940), who went on to write a series of papers on the subject in 1914, citing both Riabouchinsky and Vaschy. However, the modern form of this theorem, concerning the number of coupled variables in a phenomenon is now commonly called the Buckingham Π-theorem (chapter 5), because of the use of the Greek letter as a label for the groups of dimensionless variables.

There are several threads in the history of units and dimensional analysis. Firstly, the principle of homogeneity (i.e., not attempting to add apples and oranges—but you can multiply and divide them) was generalized from lengths to the other base dimensions; including time and mass. Joseph Fourier realized that each quantity has exponents of dimensions with respect to the fundamental, or base dimensions (length, time, etc) and that both sides of an equation must have the same exponents. For example, Fourier thought that thermal conductivity must involve, at least, length (L), mass (M), time (T) and temperature (θ), and he quite reasonably suspected that the relationship between these dimensions was a power series; that is, the dimensions of thermal conductivity, or [thermal conductivity] = $M^xL^yT^z\theta^a$. The problem is finding the values (and phases) of the exponents; this was the insight of Lord Rayleigh. This advance facilitated the use of standardized units—the Metric System, and then the SI. Finally, the technique of dimensional analysis—using dimensional arguments to infer the form of a physical relation, first developed by Lord Rayleigh was developed and formalized by other scientists.

The modern language for describing physics equations involving dimensioned quantities is a more concise and insightful manner to write physics than Galileo's style, which relied on ratios, and made a heavy use of geometry. The modern language makes possible the technique of dimensional analysis. Dimensional analysis was not invented until nearly 250 years after Galileo's *Two New Sciences*. How can this history guide us in developing a language to discuss physics? First, one should be able to write physics in manner without arbitrary choices; for example, the choice of measurement units.

However, the language should not prevent us from making these arbitrary choices; it must provide a way to translate between these dialects of the single universal language of science. When you write $d = ½\, gt^2$, where $g = 9.8$ ms^{-2}, you have made an arbitrary choice of measurement units (metres and seconds), but we know how to translate between these units and another set of units (e.g., miles and hours). Second, assign a mathematical character to a physical quantity or variable; Galileo knew that velocity had physical meaning, and Newton knew that forces had physical meaning, but they did not define them as mathematical functions that could stand alone. Today, a physicist is just as comfortable writing kg m s^{-2} (kilogram metre per second squared for force; that is, MLT^{-2}) as writing down m (metre) to represent length [7], yet there is an immense amount of physics between the two quantities.

References

[1] Fourier J-B-J 2009 *The Analytical Theory of Heat.* Cambridge University Press, (originally published 1822) (Cambridge: Cambridge University Press)

[2] Galilei G 2001 *Dialogues Concerning Two New Sciences.* Modern Library Paperback Edition (New York: Random House Inc)

[3] Newton I 1848 *Newton's Principia: The Mathematical Principles of Natural Philosophy* ed A Motte and N W Chittenden (New York: D. Adee)

[4] Williams J H 2020 *Defining and Measuring Nature: The Make of All Things* 2nd edn (Bristol: IOP Publishing)

[5] Anonymous 1805 Report on the means of measuring the initial velocity of projectiles thrown from cannon, both in an inclined and a horizontal direction. Read in the physical and mathematical class of the French National Institute in the month of December 1804 *Philosophical Magazine Series 1* **22** 220–31

[6] Rayleigh L 2011 *The Theory of Sound* in two volumes (first published 1877) (Cambridge: Cambridge University Press)

[7] Sterrett S G 2017 Physically similar systems—a history of the concept (Springer Handbook of Model-Based Science) (Berlin: Springer) 377–411

Chapter 3

Introduction to dimensions

'In order to more fully understand this reality, we must take into account other dimensions of a broader reality.'

John Archibald Wheeler (1911–2008)

As already mentioned, the dimension of any quantity, Q, may be written in the form of a dimensional product (see table 1.2 for examples),

$$\text{Dimensions of } [Q] = \mathsf{L}^{\alpha}\mathsf{M}^{\beta}\mathsf{T}^{\gamma}\mathsf{I}^{\delta}\theta^{\varepsilon}\mathsf{N}^{\zeta}\mathsf{J}^{\eta}, \tag{3.1}$$

where the exponents α, β, γ, δ, ε, ζ and η are generally small integers, they can be positive, negative or zero, and are termed dimensional exponents. This simple formula (first suggested by Fourier, and exploited by Lord Rayleigh) defines the make of all things, and is the subject of this present volume. The dimension of a physical quantity is the power with which the base dimensions that constitute that quantity is/are raised in the dimensional representation of the physical quantity of interest. We should be clear here that the dimension is not merely a power, but a combination of basic dimensions and their respective powers. Both are taken together and hence represented together. As mentioned earlier, units have a dimensional constitution; velocity has dimension of 1 in length (L) and dimension of −1 in time (T) and hence its unit is m s^{-1} (LT^{-1}). This manner of representing the dimension of a quantity was introduced by Fourier. A pair of square brackets is used to represent the dimension of a quantity with its symbol enclosed within the bracket; thus, [velocity, or v] = LT^{-1}. The dimensions of the seven base quantities of the SI are given in table 1.1; and the dimensions of a wide range of phenomena (including phenomena which are dimensionless) are given in table 1.2.

The dimensions of derived quantities include the dimensions found in the constituent base quantities. Consider force, F, defined by Newton's second law:

$F = ma$, where m is mass and a is acceleration. Thus, the dimension of force is: $[F] = [m][a]$. The dimension of mass is M, and of acceleration, represented as $[a]$, is a derived quantity being the ratio of velocity and time. In turn, velocity is also a derived quantity, being ratio of length and time. Thus,

$$[F] = \mathsf{M}[a] = \mathsf{M}([v]\mathsf{T}^{-1}) = \mathsf{M}(\mathsf{LT}^{-1}\mathsf{T}^{-1}) = \mathsf{MLT}^{-2}. \tag{3.2}$$

We read the dimension of force as: it has dimension 1 in mass, dimension 1 in length and −2 in time.

3.1 Dimensional formulae

The dimensional formula of a physical quantity is how the dimensions that constitute that quantity are represented. (It is conventional to omit power of 1.) Thus, we write: force, $[F] = \mathsf{MLT}^{-2}$; velocity, $[v] = \mathsf{LT}^{-1}$; charge, $[q] = \mathsf{IT}$; specific heat, $[c] = \mathsf{L}^2\mathsf{T}^{-2}\mathrm{K}^{-1} = \mathsf{L}^2\mathsf{T}^{-2}\theta^{-1}$; the gas constant, $[R] = \mathsf{ML}^2\mathsf{T}^{-2}\ \mathrm{K}^{-1}\ \mathrm{mol}^{-1} = \mathsf{ML}^2\mathsf{T}^{-2}\ \theta^{-1}\ \mathrm{N}^{-1}$.

It should be noted that dimensional representations of a physical quantity do not include the magnitude or value (n) of the physical quantity. Further, dimensional formulae do not distinguish the nature of the quantity; for example, the nature or types of force involved (electromagnetic, or gravitational) has no bearing on its dimensional representation—all forces have the same dimensionality, MLT^{-2}. It follows that variants of a physical quantity have the same dimensional representation; for example, velocity—whether instantaneous, average, relative—has the same dimensional representation, LT^{-1}. Moreover, the dimensional representations of a vector, and its scalar counterpart are the same—the dimensions of velocity and speed are the same. Further, dimensions of the difference of a physical quantity are the same as that of the physical quantity itself; that is, the dimensions of velocity and the difference of two velocities are the same.

A physical quantity need not have dimensions in any of the base quantities; as is the case, where physical quantities are equal to the ratio of quantities having the same dimensions. Take the case of an angle, which is a ratio of arc (L) and radius (L). The dimensions of such a physical quantity are zero in each of the basic quantities: angles, $[\theta] = \mathrm{M}^\circ\mathrm{L}^\circ\mathrm{T}^\circ$. But an angle does have a unit—the radian. The dimensionless Reynolds number, which is commonly found in fluid mechanics does not have a physical dimension, although it is made up from several commonly-measured variables. Finally, there are numerical constants such as trigonometric functions, π, etc, which are not dependent on physical quantities. The dimensions of such a constant are zero in each of the base dimensions. They are, therefore, called dimensionless constants: $[\pi] = \mathrm{M}^\circ\mathrm{L}^\circ\mathrm{T}^\circ$. However, there are physical constants, which appear as the constant of proportionality in physical formula. These constants have dimensions in base quantities; for example, constants such as the Gravitational constant, Boltzmann constant and the Planck constant are dimensional constants, and have units and dimension.

The basic approach to determine the dimensional formula of physical quantities is to use its defining equation. A defining equation like that of force in terms of mass and acceleration allows us to determine dimensional formula by further deconstructing the dependent quantities involved. However, this approach is cumbersome in certain cases when dependencies are complicated. In such cases, we should seek a simpler relation in some other context. Consider the magnetic flux density (B); we could determine its dimensional expression using, for example, Biot–Savart's law. But we also know that magnetic field strength appears in the expression for the Lorentz force. Given the Lorentz force, F ($\mathsf{MLT^{-2}}$) = charge (IT) × magnetic flux density (B) × velocity ($\mathsf{LT^{-1}}$), see figure 11.2, the dimensions of the magnetic flux density, $[B] = \mathsf{MLT^{-2}}.\{\mathsf{IT\ LT^{-1}}\}^{-1} = \mathsf{MT^{-2}}\ \mathrm{I}^{-1}$. Like-dimensions in different groups are operated upon by multiplication following the rule of indices $x^n x^m = x^{(m+n)}$.

3.2 Conversion from one system of units to another system of units

Physical quantities are expressed in a multitude of different units: SI, UK customary, c.g.s., etc, which exist for sound reasons. But whatever that reason, one must find a conversion factor which converts the value of a quantity in one system of units into a value in another system of units. The conversion of the magnitude of a unit is defined by the system of measurement itself; for example, 1 kilogram is defined to be equal to 1000 gm. But consider a general means of making the conversion of a unit of one kind, into a unit of another kind. This form of dimensional analysis leads to a dimensionless constant as conversion factor. The underlying principle involved here is that a given physical quantity has the same dimensional constitution, irrespective of the system of units. This was the great insight of Joseph Fourier; for example, whether we call a distance 1 inch or 2.54 cm, the length remains the same. As a physical quantity (for example Q) is the product of a numerical factor and a unit, we may write for that quantity in two different systems of units:

$$Q = n_1 u_1 = n_2 u_2. \tag{3.3}$$

The dimensional formula of the quantity does not vary between the two systems of units (1 and 2, in equation (3.3)). Let the dimensions of units in the two systems be:

$$[u_1] = \mathsf{M^{a1}L^{b1}T^{c1}}, \quad \text{and} \quad [u_2] = \mathsf{M^{a2}L^{b2}T^{c2}}. \tag{3.4}$$

Combining these equations, we have:

$$n_1(\mathsf{M^{a1}L^{b1}T^{c1}}) = n_2(\mathsf{M^{a2}L^{b2}T^{c2}}) \rightarrow n_2 = \left\{(\mathsf{M^{a1}L^{b1}T^{c1}})/(\mathsf{M^{a2}L^{b2}T^{c2}})\right\} n_1. \tag{3.5}$$

This is the formula used to convert a quantity from one system of units to another. As an example, consider the conversion of 1 J of energy (SI system) into ergs of energy (the c.g.s. system of units). Putting n_1=1 and applying equation (3.5), we have:

$$n_2 = (\mathsf{M^{a1}L^{b1}T^{c1}})/(\mathsf{M^{a2}L^{b2}T^{c2}}). \tag{3.6}$$

Now, the dimensional formula of energy is, [energy] = $\mathsf{ML^2T^{-2}}$. Hence, $a = 1$, $b = 2$ and $c = -2$. Putting values in the equation and rearranging, $n_2 = (1\ \text{kg}/1\ \text{gm})^1.\ (1\ \text{m}/1\ \text{cm})^2.\ (1\ \text{s}/1\ \text{s})^{-2}$. The basic units of the SI system are related to the corresponding basic units in the c.

g.s. system as: 1 kg = 10^3 gm, 1 m = 100 cm and 1 s = 1 s. Thus, $n_2 = (10^3 \text{ gm}/1 \text{ gm})^1 . (10^2 \text{ cm}/1 \text{ cm})^2 . (1 \text{ s}/1 \text{ s})^{-2} = 10^7$. Hence, 1 J = 10^7 erg.

3.2.1 The consequences of mixing units

Perhaps one of the costliest examples of not following established ideas on dimensional analysis, and of inadvertently mixing systems of units, occurred in space near Mars on 23 September, 1999, when the Mars Climate Orbiter satellite was lost during a manoeuvre to place it in an orbit around the Red Planet. After the long-crossing of interplanetary space, the satellite's controllers would have needed to slow down the satellite for it to safely enter an orbit about the planet. It is believed that it was this braking process which led to the satellite's loss. To slow down a body in motion requires the application of a force of the same order of magnitude as the force generated by the body's forward motion, but applied in the opposite direction.

Apparently, the force needed to slow the satellite down for it to enter a stable orbit was calculated in one set of units, but when the command was sent to the satellite to ignite the braking thrusters, it was applied in a different set of units. The two sets of software, on Earth and on the satellite hurtling towards Mars, did not realize that they were trying to communicate in different units; and instead of entering a stable orbit well above the surface of the planet, it attempted to enter an orbit much closer to the surface and disappeared. The principal cause of the disaster was traced to a thruster calibration table, in which British customary units instead of SI units had been used to define force. The navigation software expected the thruster impulse data to be expressed in newton seconds (the SI unit; that is, kg m s^{-1} with dimensions MLT^{-1} or $MLT^{-2}.T$), but the orbiter provided the values in pound-force seconds (a British/American customary unit; the pound-force is the product of one *avoirdupois* pound (exactly 0.453 592 37 kg) and the standard acceleration due to gravity, with dimensions $MLT^{-2}.T$). This confusion in systems of units caused the electric impulse to be interpreted as roughly one-fourth of its actual value.

In the SI system, the newton is the unit of force, and is equal to the force required to accelerate a mass of one kilogram at a rate of one metre per second squared. The pound-force (lb_f from pound with a symbol for force) is equal to the force exerted on a mass of one *avoirdupois* pound on the surface of Earth. The acceleration of the standard gravitational field (g) and the international *avoirdupois* pound (lb_m) define the pound-force as 1 lb_f = 1 $lb_m \times g$ = 1 $lb_m \times$ 32.174 049 feet per second squared, which on converting to the SI is equal to 0.454 kilogram × 9.806 65 metre per second squared = 4.448 newtons. (This is the factor which caused the loss of the satellite—it was not taken into account.)

The disappearance of the satellite illustrates an important point; even after more than two hundred years of the decimal Metric System, the world of science and technology is full of different systems of units, and converting between them requires attention.

3.3 Dimensional homogeneity

We are all familiar with the saying, you cannot add apples and oranges. This is actually a simplified expression of a far more global and fundamental rule for equations; dimensional homogeneity, stated as every additive term in an equation must have the same dimensions. Consider, for example, the change in total energy of a compressible, closed system going from state 1 to another equivalent state, 2. The change in total energy of the system (ΔE) is given by:

$$\Delta E = \Delta U + \Delta KE + \Delta PE, \tag{3.7}$$

where E has three components: internal energy (U), kinetic energy (KE) and potential energy (PE). These components can be written in terms of the: system mass (m); measurable quantities and thermodynamic properties at each of the two states, such as speed (v), elevation (z) and specific internal energy (u); and the known gravitational acceleration constant (g). Thus, $\Delta U = m(u_2 - u_1)$; $\Delta KE = ½m(v_2{}^2 - v_1{}^2)$; $\Delta PE = mg(z_2 - z_1)$.

It is straightforward to verify that the LHS of equation (3.7), and all three additive terms on the RHS of equation (3.7) have the same dimensions; that of energy ($\mathsf{ML^2T^{-2}}$):

$$[\Delta E] = [\text{Energy}] = [\text{Force. Length}] \rightarrow [\mathsf{MLT^{-2}\ L}] = [\mathsf{ML^2T^{-2}}],$$

$$[\Delta U] = [\text{mass.}\ \tfrac{\text{energy}}{\text{mass}}\ ] = [\text{energy}] \rightarrow [\mathsf{ML^2T^{-2}}],$$

$$[\Delta KE] = [\mathsf{ML^2T^{-2}}],\ \text{and}$$

$$[\Delta PE] = [\mathsf{MLT^{-2}\ L}] = [\mathsf{ML^2T^{-2}}].$$

If at some stage of an analysis, we find ourselves in a situation where two additive terms in an equation have different dimensions, this is a clear indication that we have made an error at an earlier stage in the analysis. In addition to dimensional homogeneity, calculations are valid only when the units are also homogeneous in each additive term. For example, units of energy in the above terms may be J, N m, or kg m^2 s^{-2}, all of which are equivalent. Suppose, however, that kilojoules were used in place of J for one of the terms. This term would be off by a factor of 1000 compared to the other terms.

Consider a more complex example of dimensional homogeneity—the Bernoulli equation. In fluid mechanics, the Bernoulli principle states that an increase in the speed of a fluid occurs simultaneously with a decrease in static pressure, or a decrease in the fluid's potential energy. The principle is named after the Swiss *savant* Daniel Bernoulli who published it in his *Hydrodynamica* in 1738. Although Bernoulli deduced that pressure decreases when the flow speed increases, it was Leonhard Euler who in 1752 derived Bernoulli's equation in its usual form. The principle is only applicable for isentropic flows: when the effects of irreversible processes (such as turbulence) and non-adiabatic processes (such as heat radiation) are small and can be neglected.

A common form of Bernoulli's equation, valid at any arbitrary point along a streamline, is:

$$½v^2 + gz + \frac{p}{\rho} = \text{constant}, \tag{3.8}$$

where: v is the fluid velocity at a point on a streamline, g is the acceleration due to gravity, z is the elevation of the point above a reference plane, with the positive z-direction pointing upward—so in the direction opposite to the gravitational acceleration, p is the pressure at the chosen point, and ρ is the density of the fluid at all points in the fluid. To demonstrate dimensional homogeneity, we need to verify that each additive term in the Bernoulli equation has the same dimensions; and, in addition, determine the dimensions of the constant in equation (3.8). The first term on the LHS is a velocity squared; that is, $(\mathsf{LT}^{-1})(\mathsf{LT}^{-1}) = \mathsf{L}^2\mathsf{T}^{-2}$. The second term is an acceleration multiplied by distance; that is, $(\mathsf{LT}^{-2})(\mathsf{L}) = \mathsf{L}^2\mathsf{T}^{-2}$. The third term is pressure divided by density; that is, [force/area]/[mass/volume] = $(\mathsf{MLT}^{-2}/\mathsf{L}^2)/(\mathsf{ML}^{-3}) = \mathsf{L}^2\mathsf{T}^{-2}$. Indeed, all three additive terms have the same dimensions—there is homogeneity.

Invoking dimensional homogeneity, the constant on the RHS of equation (3.8) must have the same dimensions as the other additive terms in the equation. Thus, the primary dimension of the Bernoulli constant (in this form) is $\mathsf{L}^2\mathsf{T}^{-2}$. The Bernoulli equation may be written in other ways; for example, by taking equation (3.8) and multiply throughout by the density. This will give a different set of units for the constant. But if the dimensions of any of the terms had been found to be different from the others, it would indicate that an error was made somewhere in the analysis. An additional example of dimensional homogeneity is found in section 11.3; dealing with electromagnetic quantities.

3.3.1 Checking equations for dimensional consistency

This is the manner in which most students first encounter dimensional analysis. Consider the physical quantities s, v, a and t with dimensions: $[s] = \mathsf{L}$, $[v] = \mathsf{LT}^{-1}$, $[g] = \mathsf{LT}^{-2}$ and $[t] = \mathsf{T}$; that is, a length, a velocity, an acceleration and a time duration. How would we determine whether each of the following equations is dimensionally consistent: (a) $s = vt + \frac{1}{2}gt^2$; (b) $s = vt^2 + \frac{1}{2}gt$; and (c) $v = \sin(gt^2/s)$. By the definition of dimensional consistency, we need to check that each term in a given equation has the same dimensions, and that the arguments of any standard mathematical functions are dimensionless. In (a), there are three terms, one in the LHS and two on the RHS: $[s] = \mathsf{L}$; $[vt] = [v]\,[t] = \mathsf{LT}^{-1}\,\mathsf{T} = \mathsf{L}$; and $[\frac{1}{2}gt^2] = [g]\,[t]^2 = \mathsf{LT}^{-2}\,\mathsf{T}^2 = \mathsf{L}$. All three terms have the same dimension, so this equation is dimensionally consistent.

For (b): $[s] = \mathsf{L}$; $[vt^2] = [v]\,[t]^2 = \mathsf{LT}^{-1}\,\mathsf{T}^2 = \mathsf{LT}$; and $[gt] = [g]\,[t] = \mathsf{LT}^{-2}\,\mathsf{T} = \mathsf{LT}^{-1}$. In this case, each of the equations is dimensionally unique. If the equation contains a trigonometric function, we should check that the argument of the sine function is dimensionless: $[gt^2/s] = [g]\,[t]^2/[s] = \mathsf{LT}^{-2}\,\mathsf{T}^2/\mathsf{L} = \mathsf{L}/\mathsf{L} = 1$. The argument is dimensionless. Now we need to check the dimensions of each of the two terms (LHS and RHS) in the equation: $[v] = \mathsf{LT}^{-1}$; and $[\sin(gt^2/s)] = 1$. The two terms have different dimensions—the equation is not dimensionally consistent.

3.3.2 Some details: [x] represents the physical dimension of x

One further point is the effect of the operations of calculus on dimensions. We have seen that dimensions obey the rules of algebra, just like units, but what happens when we take the derivative of one physical quantity with respect to another, or integrate a physical quantity over another? The derivative of a function is just the slope of the line tangent to its graph and slopes are ratios; so for physical quantities v and t, the dimension of the derivative of v with respect to t is the ratio of the dimension of v over that of t:

$$\left[\frac{dv}{dt}\right] = [v]/[t], \quad \text{or} \quad \partial^2 f/\partial x \partial y = f/[x][y]. \tag{3.9}$$

Similarly, since integrals are sums of products, the dimension of the integral of v with respect to t is simply the dimension of v times the dimension of t:

$$\left[\int v\mathrm{d}t\right] = [v][t], \text{ or} \left[\int_{x1}^{x2} f(x)dx\right] = [f(x)][x]. \tag{3.10}$$

Generally, the dimension of a variable is written in operator form, and is subject to certain rules:

$$\begin{aligned} &[x.\ y] = [x].\ [y]; \\ &[x/y] = [x]/[y]; \\ &[x^n] = [x]^n; \\ &[x]+/-[y] = [x] = [y]; \quad \text{and} \\ &[x/x] = 1. \end{aligned} \tag{3.11}$$

Note that if an expression involving functions as e^{at} or $\sin(at)$ where t is time appears in a model, then a must have dimensions T^{-1}, for the argument to be without dimension Also, recall that the dimension of the derivative is the ratio of the dimensions; for example, the derivative of pressure (p) with respect to time (t) has the units: $\left[\frac{dp}{dt}\right] = [p]/[t] = \mathsf{ML^{-1}T^{-2}/T} = \mathsf{ML^{-1}\ T^{-3}}$, and a new derived unit is generated. An analogous rule applies to the partial derivatives.

3.4 Approaching dimensional analysis

The principle of dimensional analysis is based on the idea that only similar quantities can be added, subtracted and equated, and follows from the principle of homogeneity. Dimensional analysis gives an insight into the composition of a physical quantity. Take the case of power; the SI unit is the watt (kg $\mathrm{m}^2\ \mathrm{s}^{-3}$). What is its relation with the appropriate base quantities? Notwithstanding the exact definition, its dimensional formula tells us it is $\mathsf{ML^2T^{-3}}$; that is, it has the dimension of 1 in mass, 2 in length and 3 in time. An elaborate form of the homogeneity principle is known as the Buckingham Π-theorem (see chapter 5), states that an expression of physical quantity having n variables can be reduced to an expression of n-m nondimensional parameters; where m is the numbers of base dimensions involved (three in the case of power). We will see later that this principle may be widely used to establish equations of physical quantities.

Not surprisingly, any equation derived during an investigation of a phenomenon should be dimensionally consistent. This necessity can be a surprisingly powerful analytical tool. Firstly, it provides a way to quickly check as to whether or not an answer is correct. Moreover, there are certain problems that can be answered using such dimensional analysis alone; allowing one to avoid calculations all together.

Many physical quantities are expressed in terms of transcendental functions such as trigonometric, exponential or logarithmic functions. An immediate consequence of the homogeneity principle is that we can only have dimensionless groups as the argument of these transcendental functions. Such functions are expansions involving power terms:

$$e^x = 1 + \frac{x}{1!} + \frac{x^2}{2!} + \cdots$$

$$\sin x = x - \frac{x^3}{3!} + \frac{x^5}{5!} - \cdots, \text{ and}$$

$$\tan x = x + \frac{x^3}{3} + \frac{2x^5}{15} + \cdots$$

In all these expansions, if x is a dimensional quantity, then each term on the RHS will have different dimensions as x is raised to different powers. This violates the principle of homogeneity. Angle being dimensionless is permissible here, as no dimensions are involved. However, when we have expressions involving physical quantities with transcendental functions, in accordance with the principle of homogeneity, the argument to the functions needs to be dimensionless; for example, in the wave equation: $y = A\sin(kx - \omega t)$ both kx and ωt need to be dimensionless. If, however, x was a dimensionless quantity (for example, a Reynolds number—see next chapter for details), then x^2, x^3, etc, would all be without dimension.

3.4.1 Checking a formula

The terms of an equation connecting different physical quantities should be dimensionally compatible. In accordance with the principle of homogeneity, each term of the equation has the same dimensions. Hence, a formula not having dimensionally compatible terms is incorrect. We seek the situation where: (*a*) dimensions on both sides of an equation are equal, and (*b*) terms connected with plus or minus signs in the expression have the same dimensions.

Consider, for example, the Van der Waals equation of state which defines departures from the ideal behaviour of gases; that is, deviations from $PV = RT$. We may write for the van der Waals equation:

$$(P + a/V^2)(V - b) = RT, \tag{3.12}$$

where P is pressure, V is volume, T is temperature and R is the gas constant. But what are the units of the constants a and b? Applying the principle of homogeneity of dimensions, we see that the dimensions of the added and subtracted terms are equal. Hence,

$$[a/V^2] = [P] = \left[\frac{\text{Force}}{\text{Area}}\right] = (\text{MLT}^{-2}/L^2) = \text{ML}^{-1}\text{T}^{-2}$$
$$\rightarrow [a] = (\text{ML}^{-1}\text{T}^{-2})[V^2] = (\text{ML}^{-1}\text{T}^{-2})(\text{L}^3)^2 = \text{ML}^5\text{T}^{-2}. \tag{3.13}$$

In the SI, the unit of a would be kg m^5 s^{-2}. Also, $[b] = [V] = \text{L}^3$, and the unit of b is identical to that of volume, m^3.

3.4.2 Deriving a formula

This is, perhaps, the most curious aspect of dimensional analysis. If we had a general facility to construct formulae in this manner, we would readily discover a great many more of the secrets of Nature. Clearly, derivation of formulae by dimensional analysis is only possible under certain circumstances. Even then, the significance of deriving formulae is important, as it has contributed quite remarkably to the development of fluid mechanics, and in other branches of the physical and biological sciences. Here, we will look briefly at the formation of drops (this will be examined later in more detail).

We consider the interface between two fluids; for example, water and air. In particular, we will examine the relative magnitude of the fluid's inertia compared to its surface tension. This ratio is useful in analysing thin film flows, and the formation of droplets and bubbles. Let us first consider the variables or quantities that will be of importance in the formation of a droplet. Size will be important (so there is a length scale, r); then the fluid under consideration must be considered—the fluid's density (ρ) and its surface tension (σ). The surface tension is a measure of how strongly the molecules that compose the fluid are interacting. Due to attractive intermolecular forces, a molecule in the fluid is pulled equally in every direction by neighbouring liquid molecules, resulting in a net force of zero. The molecules at the surface, however, do not have the same density of molecules on all sides of them and therefore are pulled inward. This creates some internal pressure and forces liquid surfaces to contract to the minimum area. The other quantity involved in drop formation is the velocity of the fluid phase (v).

There are two forces involved: a measure of the fluid's inertia, F_A, and a measure of the cohesive force generating a surface, F_coh. These forces have the same dimensions; thus, $[F_\text{A}] = [F_\text{coh}] = \text{MLT}^{-2}$. The dimensions of the variables mentioned above are given in table 3.1.

Table 3.1. Variables for drop formation.

Variable	Dimension	Description
r	L	Length scale of droplet
ρ	ML^{-3}	Density of fluid
σ	MT^{-2}	Surface tension of fluid: force per unit length, or MLT^{-2}/L
v	LT^{-1}	Velocity of fluid

Given the dimensions in table 3.1, we can derive an expression to describe the ratio of the two forces (F_A and F_{coh}) that defines how readily a droplet will form at an interface. This dimensionless ratio of two forces, We, will be: We $\propto$ (r, ρ, σ, v) = constant (r, ρ, σ, v); that is, it will be a function of all the selected variables. And the problem is to identify the dimensionless combination of variables that is equal to the dimensionless ratio, We. This can be done by multiplying and dividing the variables, and powers of the variables, until we generate a dimensionless combination. For example, we must remove the mass dependence of We by dividing ρ by σ, and then using powers of r to eliminate the length dependence. There are rules to do this in dimensional analysis, and these rules will be discussed in detail in the coming chapters. But in the present case, the dimensionless combination of all the variables is:

$$\mathrm{We} = rv^2\rho/\sigma. \tag{3.14}$$

We can check the dimensions of this ratio as: [We] = $\mathsf{L}\ (\mathsf{LT}^{-1})^2\ \mathsf{ML}^{-3}/\mathsf{MT}^{-2} = 1$. We could also write this ratio of two forces (each with the dimension MLT^{-2}) as: $\mathrm{We} = F_A/F_{coh} = r^2\rho v^2/r\sigma = rv^2\rho/\sigma$.

Of course, as we are dealing with a dimensionless ratio we could equally well have written: $\mathrm{We} = \sigma/rv^2\rho$. How do we distinguish between the two possibilities? As with the reality or otherwise of all theoretical models, the acid test is comparison with experiment; for example, measure the dynamical behaviour (observe the formation of drops) between two very different interfaces: water and air (σ = 72.86 mN m^{-1} at 293 K, with a density of 1 g cm^{-3}), and propylene carbonate and air (σ = 2.9 mN m^{-1} at 293 K, with a density of 1.2 g cm^{-3}). A measurement of We for these two systems, should confirm the validity of equation (3.14).

This example highlights the power and the limitations of dimensional analysis. First, we have created a model based on a hypothesis; a model which may be verified by experiment. But we also see that it is not possible to evaluate numerical constants by dimensional analysis. It can also be seen that dimensional analysis cannot be applied in a situation where there are sums or differences of terms, as in the case of $x = ut + \frac{1}{2}gt^2$. We cannot derive this equation, as dimensional analysis will yield one of the two terms—not both simultaneously.

Further reading

There are sections on dimensional analysis in many of the larger textbooks of fluid mechanics. However, these sources tend to concentrate on the application of dimensional analysis to that field. An excellent introductory text on the subject of dimensions of quantities and dimensional analysis is: Bridgman P W 1922 *Dimensional Analysis* (New Haven, CT: Yale University Press) This is one of the first texts on dimensional analysis, and is a classic. Despite its age this book has been reissued several times and is readily available.

Chapter 4

Why, and how we play with variables

'There must be as many different units as there are different kinds of quantities to be measured, but in all dynamical sciences it is possible to define these units in terms of the three fundamental units of Length, Time, and Mass'.

James Clerk Maxwell

The purpose of establishing the physical relationship between a set of observable quantities (that is, the purpose of physics) is to find an equation that relates the variables of the system under study—whether it be a rocket trying to escape the Earth's gravitational field, water flowing through the vascular system of a plant, or observing the temporal evolution of a nuclear explosion as the fireball expands against the density of air. Such defining relationships are useful because they capture the behaviour of many quantities (temperature, air pressure, air density, gravity, intermolecular forces, etc) in a single, comprehensive statement of equality; that is, a defining equation. For example, we observe via time-lapse photography how the atomic bomb's blast radius grows over time, and given the density of the atmosphere we may calculate the bomb's explosive power. We can even estimate the amount of time needed to cook a large turkey (see section 15.1), or to relate the intermolecular vibrations in a solid to the melting point of that solid (see section 12.4). Because such equalities are so useful, a great deal of time and effort is devoted to establishing them through mathematical techniques, experimental observation, and rules of thumb.

But before we move on, it is useful to introduce two symbols that do not, however, define, or express equivalence or equality, but are nevertheless essential in physics, and especially in dimensional analysis. There is the tilde, $\sim$, and the symbol for proportionality, $\propto$. If a quantity g is equal to some combination of factors, $f(r_1, \ldots, r_n)$, save for some numerical multiplier, or constant γ ; such that, $g = \gamma f(r_1,\ldots, r_n)$, then we say $g \sim f(r_1,\ldots, r_n)$. For example, if $f = 24\pi l^2$, or $f = 100\ l^2$, we say in both cases that $f \sim l^2$, for all f for which this relation holds; that is, f scales as l^2. Up to some numerical

doi:10.1088/978-0-7503-3655-0ch4

constant that can be obtained by measuring f at values of l^2, f is determined entirely through its dependence on the quantity l. Another closely related notation is $\propto$, which is used to denote that the LHS of an equation is proportional to the RHS. In contrast to $x \sim l$, $x \propto l$ does not mean that the behaviour of x is captured solely through l, but that l is a factor in the definition of x. For instance, if $f(x,y,z) = \Pi_{xyz}$, we can say: $f \propto x$, $f \propto y$, and $f \propto z$ simultaneously.

As we mentioned earlier, any true equation expressing physical quantities must have the same units on both sides. Thus, if we ignore dimensionless pre-factors (pure numbers), we have LHS $\sim$ RHS, or equivalently, RHS/LHS ~ 1 for all physical relations. More generally, we can replace RHS/LHS by an arbitrary product containing all the problem's relevant variables raised to as yet unspecified powers; i.e. $x^\alpha y^\beta z^\gamma \sim 1$. We call this product a scaling relation, because it shows how each variable will change as the scale (magnitude) of the other variables are changed. Such a scaling relation is composed of different factors, each of which has possibly different combinations of the base dimensions involved in the phenomenon under investigation. Suppose we have a system whose relevant variables are known to be position, r, or length, L, velocity, v, or length per unit time, and time t, so that our scaling relation is given by $r^\alpha v^\beta t^\gamma \sim 1$. We denote the dimensions of the variable x by $[x]$, and we have $[r] = \mathsf{L}$, $[v] = \mathsf{LT}^{-1}$ and $[t] = \mathsf{T}$. If we replace our variables by their base dimensions, we obtain a dimensional equation: $\mathsf{L}^\alpha\,(\mathsf{LT}^{-1})^\beta \mathsf{T}^\gamma \sim 1$, or $\mathsf{L}^{\alpha+\beta}\,\mathsf{T}^{-\beta+\gamma} \sim 1 = \mathsf{L}^\circ \mathsf{T}^\circ$. Thus, we find that the two equations $\alpha+\beta = 0$ and $-\beta+\gamma = 0$ govern the behaviour of our scaling relation so that $-\alpha = \beta = \gamma$. Choosing $\alpha = -1$, we have $\beta = \gamma = 1$; so that our dimensionless relation becomes $r^{-1}vt \sim 1$, which leads to $r \sim vt$, the usual relation between position and velocity.

Thus, without knowing anything about kinematics, we were able to obtain a defining relation solely using dimensional analysis. Had we chosen another value for α, say 5/6, we would still end up with the same scaling relation since we would have taken the $\sqrt[5/6]{r}$ to obtain an integer power for a variable of interest. The value we choose is inconsequential; the scaling conditions demand certain relative values for the exponents, and it is their relative values that govern the system. The technique might not yet seem impressive, and $r = vt$ is, perhaps, a trivial result. However, dimensional analysis is useful far beyond such a simple example and can produce results in situations where intuition fails. All that is required of the user is to determine which physical quantities or variables should be important, and unimportant for the phenomenon under consideration.

If there are n variables describing a physical system, with m independent dimensions (for example; M, L, T and I), the variables will form $l = n-m$ independent dimensionless parameters ($\Pi_1,\ldots, \Pi_l$) related by $\Pi_1 = f(\Pi_2,\ldots, \Pi_l)$. This is the statement of the Buckingham Π-theorem, which we will look at in some detail shortly. It is a formula for generating dimensionless groups of variables, or groups of variables with a particular character (see chapter 5). But for the moment, we will restrict ourselves to the manipulation of power-law dependencies that is at the heart of the Buckingham Π-theorem (often termed the Rayleigh algorithm). So far, our procedure has been to identify n physical variables $a,b,c,\ldots$, form a scaling product $a^\alpha \times b^\beta \times c^\delta \times \cdots$, and substitute the m independent dimensions of our variables to form a system of m equations in n variables. Solving these equations yields the exponent values needed to form a dimensionless scaling product, $\Pi \sim 1$.

This procedure yields unique solutions that lead to a single dimensionless product. In general, however, such a system will yield a total of $n-m$ independent solutions, so we can have multiple dimensionless scaling products $\{\Pi_1, \Pi_2 \ldots\}$. In such cases, the analysis no longer implies $\Pi_i \sim 1$, but instead $g(\Pi_1) = f(\Pi_2)$, where f and g are arbitrary functions. For simplicity, we can say that we have $\Pi_1 = f(\Pi_2)$, and in general $\Pi_1 = f(\Pi_2, \ldots, \Pi_n)$.

We will see that the technique is straightforward to use, and can provide physical insight. With its simplicity, however, comes a number of shortcomings. Although we can usually get to the scaling solution for a problem, dimensional analysis can never yield exact answers in cases of numerical factors—you always lose the 4π from the $4\pi\varepsilon_0$, because numbers are dimensionless. In cases where we end up with multiple dimensionless products $\{\Pi_1, \Pi_2, \ldots\}$ such that $\Pi_1 \sim f(\Pi_2, \ldots)$, we need to come up with physical reasons to narrow down the form of f, and take limiting cases to demonstrate the plausibility of a proposed solution. However, as one of the reasons for using dimensional analysis is because the problem under study is complex, this can be a tall order. Moreover, the method works best when the answer is a simple product of variables. When the answer involves the sum of variables of the same dimension, things become difficult. In these cases, we may be able to find several limiting cases (when one, or the other summand goes to zero) and then combine the solutions afterwards. However, both as a learning and a research tool, dimensional analysis is a powerful first, and last line of attack on new problems. But for the unwary student, dimensional analysis can become a deep rabbit home in which to lose oneself.

4.1 The de-dimensionalization of equations

Dimensional homogeneity guarantees that every additive term in an equation has the same dimension. It follows that if we divide each term in the equation by a collection of variables and constants whose product has those same dimensions, the equation may be rendered nondimensional. If, in addition, the nondimensional terms in the equation are of order unity, the equation is called normalized. In the process of de-dimensionalizing an equation of motion, nondimensional parameters often appear—most of which are named after a notable scientist or engineer (see section 4.2). Consider the equation of motion describing the elevation z of an object falling by gravity through a vacuum (no air drag). The initial location of the object is z_0, and its initial velocity is w_0 in the z-direction. We may write for the equation of motion:

$$(\mathrm{d}^2 z/\mathrm{d}t^2) = -g. \tag{4.1}$$

For the simple differential equation given in equation (4.1) there are two dimensional variables: z (of dimension L) and t (of dimension T). Nondimensional (dimensionless) variables are defined as quantities that change or vary in the problem, but are without dimensions; for example, angle of rotation measured in degrees, or radians. The gravitational constant, g, while dimensional, remains constant and is called a dimensional constant. Two additional dimensional constants are relevant to this particular problem: initial location z_0, and initial vertical speed w_0. Equation (4.1) is easily solved by integrating twice, and applying the initial conditions. The result is an expression for elevation, z at any time t:

$$z = z_0 + w_o t - \tfrac{1}{2} g t^2. \tag{4.2}$$

The constant of ½ and the exponent 2 in equation (4.2) are results of the integration. To de-dimensionalize equation (4.1), we need to select scaling parameters, based on the primary dimensions contained in the original equation. In problems related to fluid flow, for example, there are typically three scaling parameters, since there are usually three primary dimensions in such problems. The scaling parameters usually include a characteristic length, L, a characteristic velocity, v, and a reference pressure difference, P_0—P_∞. Other parameters and fluid properties such as density (ρ), viscosity (μ) and gravitational acceleration (g) may also enter the problem (as we will see later).

In the case of a falling object, there are only two primary dimensions, length and time, and we are limited to selecting only two scaling parameters. We have some options in the selection of the scaling parameters since we have three available dimensional constants: g, z_0 and w_0. Having chosen z_0 and w_0 as scaling parameters, we de-dimensionalize the dimensional variables z and t. The first step is to list the primary dimensions of all dimensional variables and dimensional constants in the problem. The primary, or base dimensions of all parameters are: $[z]$ = L; $[t]$ = T; $[z_o]$ = L; $[w_o]$ = L/T and $[g]$ = L/T^2.

The second step is to use our two scaling parameters to de-dimensionalize z and t (by inspection) into nondimensional variables z^* and t^*:

$$\text{de-dimensionalized variables: } z^* = z/z_0;\ t^* = w_0 t/z_0 \tag{4.3}$$

Substituting the nondimensional variables into equation (4.1) gives,

$$\begin{aligned} &(\mathrm{d}^2 z/\mathrm{d}t^2) = (\mathrm{d}^2(z_o z^*)/\mathrm{d}(z_o t^*/w_o)^2) = (w_o^2/z_0)(\mathrm{d}^2 z^*/\mathrm{d}t^{*2}) = -g \\ &\rightarrow (w_o^2/g z_0)(\mathrm{d}^2 z^*/\mathrm{d}t^{*2}) = -1, \end{aligned} \tag{4.4}$$

which is the desired nondimensional equation. The grouping of dimensional constants in equation (4.4) is the square of a well-known nondimensional parameter, or dimensionless group known as the Froude number; Froude number: $\mathrm{Fr} = w_0/\sqrt{g z_0}$. The reader will discover that a great many problems in dimensional analysis, particularly in fluids, resolve themselves into generating well-known nondimensional groups.

William Froude (1810–1879) was an English engineer, hydrodynamicist and naval architect. He was the first to formulate reliable laws for the resistance that water offers to ships. In continuum mechanics, the Froude number (Fr) is a dimensionless number defined as the ratio of the flow inertia to the external field (the latter in many applications is simply that due to gravity). It is based on the speed–length ratio which he defined as: $\mathrm{Fr} = v/\sqrt{gL}$, where v is the local flow velocity, g is the local external field (acceleration due to gravity) and L is a characteristic length.

We can now write equation (4.4) in terms of the nondimensional parameter, the Froude number,

$$(\mathrm{d}^2z^*/\mathrm{d}t^{*2}) = -1/\mathrm{Fr}^2. \quad (4.5)$$

In its dimensionless form, only one parameter remains. Equation (4.5) (like equation (4.1)) can be readily solved by integrating twice, and applying the initial conditions. The result is an expression for dimensionless elevation, z^* at any dimensionless time, t^*:

$$z^* = 1 + t^* - (-1/\mathrm{Fr}^2)t^{*2} \quad (4.6)$$

Comparison of equation (4.2) and equation (4.6) reveals them to be equivalent. What then is the advantage of de-dimensionalizing the equation? Before answering, we note that the advantages are not so clear in this simple example, because we were able to analytically integrate the differential equation of motion. In more complicated problems, the differential equation (or more generally the coupled set of differential equations) cannot be integrated analytically, and we must either integrate the equations numerically, or design and conduct experiments to obtain the needed results, both of which can incur considerable time and expense. In such cases, the nondimensional parameters generated by de-dimensionalizing the equations are extremely useful and can save a great deal of time and effort.

There are two key advantages of de-dimensionalization. First, it increases our insight about the relationships between key parameters. The form of the Froude number reveals, for example, that doubling w_0 has the same effect as decreasing z_0 by a factor of 4. Second, it reduces the number of parameters in the problem; for example, the original problem contains one dependent variable, z; one independent variable, t; and three additional dimensional constants, g, w_0 and z_0. The de-dimensionalized problem contains one dependent parameter, z^*; one independent parameter, t^*; and only one additional parameter, namely the dimensionless Froude number, Fr. The number of additional parameters has been reduced from three to one.

If one wished to explore the experimental consequences of equation (4.1) in the laboratory, for example by dropping steel balls in tubes; without invoking non-dimensional parameters, one would need to keep certain parameters fixed, while the trajectory of the steel ball was observed. In order to reasonably document the trajectories for a range of all three of the dimensional parameters g, z_0 and w_0, this brute force method would require several (a minimum of four) additional plots at various values of w_0; plus, several additional sets of such plots for a range of g. A complete data set for three parameters with five levels of each parameter would require 5^3 or 125 experiments. De-dimensionalization reduces the number of parameters from three to one—a total of only five experiments are required to achieve the same experimental resolution [1].

Another advantage of de-dimensionalization is that extrapolation to untested values of one or more of the dimensional parameters becomes possible—we have a model, which may be explored. For example, suppose the experiment were made at only one value of gravitational acceleration, but you wished to extrapolate these data to a different value of g. This is easily accomplished via the dimensionless data. Thus, experimental data obtained on Earth may also be used to predict the behaviour on another body—an example of the great principle of similitude.

4.2 Some of the more widely-used nondimensional groups

In a particular model system, dynamic similarity (similitude) requires that the ratio of all forces be the same. The ratio of different forces produces many of the key nondimensional parameters found in fluid mechanics. (Note that 'inertial force' means mass × acceleration; as it is equal to the total applied force, it is often one of the two forces in a ratio.) Table 4.1 contains come commonly encountered nondimensional numbers or quantities.

These nondimensional groups occur regularly when dimensional analysis is applied to fluid-dynamical problems (as we will see in this volume). They can be derived by considering the forces acting on a small volume of fluid. They can also be derived by de-dimensionalizing the differential equations of fluid flow [1]. But when using them, one also has to take care with systems of units. In the SI system of units, the Reynolds number involves: fluid flow (in m s^{-1}), a length-scale (in m), fluid density (in kg m^{-3}) and viscosity (kg m^{-1} s^{-1}).

Finally, the dimensionless combinations that you can make in a given problem are not unique: if x and y are both dimensionless, then so are xy and x^2y and $x + y$ and, indeed, any function that you want to make out of these two variables. There are other reasons to be interested in dimensionless quantities. The first is practical: identifying dimensionless quantities at an early stage in a calculation will save you time in the subsequent analysis. In a calculation involving many variables, one often finds the same dimensionless combinations of variables appearing at every stage; for example, the electronic charge, e and the permittivity of free space, ε_0. In particular, it is only dimensionless combinations that can appear as the arguments of functions. The second reason to be interested in dimensionless quantities is because the answer to a calculation often simplifies in certain regimes. Perhaps this is the regime of long-time, or short-distance, or high-speed. But only dimensionless numbers can be big, or small. For a dimensionless quantity x, we can write $x >> 1$. But it makes no sense to write $Y >> 1$ if Y is not dimensionless: a dimensionful quantity (the antonym of dimensionless) must always be big or small relative to something else.

4.3 Some examples of straightforward dimensional analyses

4.3.1 The Trinity explosion

Sir Geoffrey Ingram Taylor (1886–1975) was a British physicist and mathematician, and a major figure in fluid dynamics and wave theory. Indeed, Taylor's name is

Table 4.1. Some commonly encountered nondimensional groups, or numbers. Here, flow velocity, v; length-scale, L; viscosity, μ; surface tension, σ; speed of sound, c; pressure, p; density, ρ; acceleration due to gravity, g; angular frequency, Ω; specific heat, cp; thermal conductivity, k.

Nondimensional group	Description	Application
Reynolds number, Re	$\frac{\rho v L}{\mu} = \frac{\text{inertial force}}{\text{viscous force}}$	Viscous flows
Froude number, Fr	$\frac{v}{\sqrt{gL}} = (\frac{\text{inertial force}}{\text{gravitational force}})^{1/2}$	Free-surface flows
Weber number, We	$(\rho v^2 L/\sigma) = \frac{\text{inertial force}}{\text{surface tension}}$	Surface tension and surface instability (see section 4.3.4)
Rossby number, Ro	$\frac{v}{\Omega L} = \frac{\text{inertial force}}{\text{Coriolis force}}$	Rotating flows
Mach number, Ma	$\frac{v}{c} = (\frac{\text{inertial force}}{\text{compressibility force}})^{1/2}$	Compressible flows
Prandtl number, Pr	$\frac{\mu cp}{k} = \frac{\text{Momentum diffusivity}}{\text{Thermal diffusivity}}$	Heat transfer in flows
Euler number, Eu	$(\rho v^2/\Delta p) = \frac{\text{dynamic pressure}}{\text{pressure drop or difference}}$	Aerodynamics

associated with 21 different, fundamental processes in fluid mechanics[1]. He is prominent in the field of constructing simple models to describe complex phenomena, so as to obtain theories that may be examined and verified experimentally. Here, we look at something Taylor saw, and then went on to describe in so much detail that he was considered a Cold War security risk.

Geoffrey Taylor was present at the Trinity (nuclear test) explosion, July 16, 1945, as part of General Leslie Groves' VIP list of ten people who observed the test explosion from Compania Hill, about 20 miles northwest of the tower. In 1950, Taylor published two papers [2] estimating the yield of that explosion using dimensional analysis of high-speed photography stills from that test, bearing timestamps and a physical scale of the blast radius that had been published in *Life* magazine (see figure 4.1 for other examples). His final theoretical estimate of 22 kilotons of TNT for the yield of the explosion was remarkably close to the accepted value of 20 kilotons that was still classified at that time. Merely by looking at a series of time-lapse photographs of the explosion, Taylor made a simple dimensional analysis of the expanding fireball to estimate the quantity of energy released. His analysis is one of the most famous, and certainly one of the simplest to use dimensional analysis to productively model a hugely complex problem—where direct experimental investigation is not possible.

To analyse the images shown in figure 4.1, Taylor made two assumptions [2]:

- The total energy (E) was released in a short period of time, quasi instantaneously, and
- The shock wave was spherical.

The published photographs revealed the time evolution of the size of the fireball, of radius, r; that is, the photographs provided r as a function of time, t. The question of interest was the reality, or otherwise of the hypothesis:

$$r \propto (E)(t)(\rho), \tag{4.7}$$

where E is the energy output—the 'tonnage' of the explosion, t is the times after the explosion and ρ is the density of the air into which the fireball is expanding. The dimensions of these variables are given in table 4.2.

From the initial hypothesis, we can invoke the Rayleigh algorithm and assume that the variables are involved in the determination of r as a power series; that is, we may write,

$$[r] = \mathsf{L} = [E]^x[\rho]^y[t]^z. \tag{4.8}$$

Substituting the dimensions of the variables, we may write the dimensional equation as:

$$[r] = \mathsf{L} = (\mathsf{ML}^2\mathsf{T}^{-2})^x(\mathsf{ML}^{-3})^y(\mathsf{T})^z = \mathsf{M}^{x+y}\mathsf{L}^{2x-3y}\mathsf{T}^{-2x+z}. \tag{4.9}$$

[1] G I Taylor—Wikipedia.

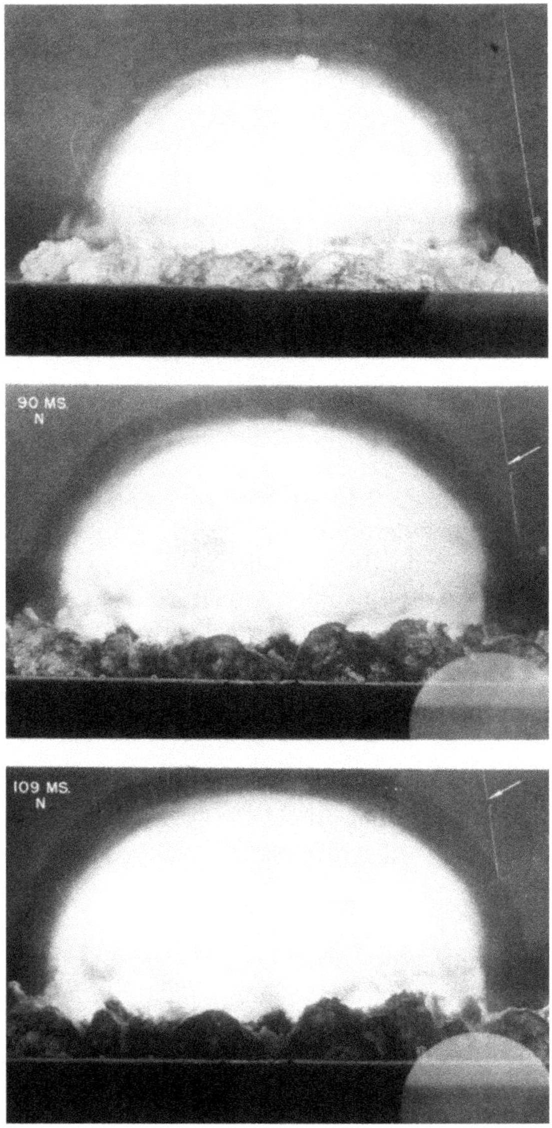

Figure 4.1. The Trinity Test, July 16, 1945, Trinity Site Zero, Alamogordo Test Range, USA. The expanding fireball. 'If the radiance of a thousand suns, Were to burst at once into the sky, That would be like the splendour of the Mighty One… I am become Death, The destroyer of Worlds.' The moment Lord Krishna shows his universal form to Prince Arjuna, Bhagavad-Gita. Reproduced from Wikipedia Trinity (nuclear test) by Atomic Photographers Guild Berlyn Brixner. Image stated to be in the public domain. It is included within this article on that basis.

M is to the $x + y$ power, because energy and density are both dependent on M. L is to the $2x - 3y$ power, because energy is dependent on the square of distance, and density is dependent on one over the cube of distance. T is to the $-2x + z$ power, because energy is dependent on one over the square of time, and time is dependent on time. This provides three simultaneous equations:

Table 4.2. Variables for a nuclear explosion.

Variable	Dimension	Description
r	L	Expanding radius of fireball
E	ML^2T^{-2}	Energy released at $t = 0$
t	T	Time after explosion
ρ	ML^{-3}	Density of air

$$\text{M}: x + y = 0$$

$$\text{L}: 2x - 3y = 1$$

$$\text{T}: -2x + z = 0,$$

yielding the results: $x = 1/5$, $y = -1/5$, $z = 2/5$.

The radius of the shock wave is therefore:

$$r = C\{E^{1/5}\rho^{-1/5}t^{2/5}\}, \tag{4.10}$$

where C is a constant of proportionality, which we will assume is of order 1. Solving the equation for E we obtain for the energy released by the explosion: $E = (r^5\rho)/t^2$.

Geoffrey Taylor analysed the time-lapse photographs of the expanding fireball (see figure 4.1), and saw that at t =.006 seconds, the shock wave was approximately 80 metres in diameter. He knew that the density of air is $\rho = 1.2$ kg m^{-3}. Putting these values into the energy equation gives: $E = 1 \times 10^{14}$ J, which is equivalent to about 17 kilotons of TNT. Geoffrey Taylor subsequently refined his model of the explosion [2], by looking at the time evolution of the radius of the expanding fireball. Taylor used data from the series of photographs to plot the graph of log r against log t (by analysing the evolution of the fireball from many photographs, he was able to reduce the overall uncertainty in the calculation). As $r = (Et^2/\rho)^{1/5}$, a graph of (5/2)log r vs log t gives a straight line with an intercept of ½log(E/ρ), from which E can then be robustly estimated.

4.3.1.1 Postscript: the Beirut explosion

The terrible explosion in the port facility of Beirut, Lebanon, on 4th August 2020, was caused by the detonation of a warehouse full of ammonium nitrate (NH_4NO_3—lots of hydrogen and oxygen atoms in close proximity, which combine vigorously (exothermically) to form water, releasing a great deal of energy, and a gas, which expands rapidly). Interestingly, it is possible to calculate a good estimate of the total energy yield of the explosion from mobile-phone videos of the event. The analysis of the explosion in Beirut follows the analysis by Geoffrey Taylor of the Trinity atomic bomb test. Assume that a large amount of energy E is deposited instantly in a tiny volume at time $t = 0$, and this produces a shock wave that expands spherically with a time-dependent radius $R(t)$ into the surrounding air of mass density ρ.

There are many videos of the explosion on YouTube, and from these images you may readily estimate that the shock wave had a diameter of about 600 m after about 1 s. Then using 1.2 kg m^{-3} for the density of air, one calculates an estimated (explosive) yield of about 3×10^9 J, or the equivalent of around 720 tons of TNT. This estimate is in reasonable agreement with the known facts of the explosion: that is, that it resulted from the detonation of 2750 tons of ammonium nitrate. However, one should take care when you are using variables raised to the 5th power—a small error in the determination of the variable (for example, scanning by eye a photograph with a ruler) is greatly magnified when you take its fifth power.

4.3.2 The smallest measurement-scale

I am sure, we have all at some time observed how systems of units can be confusing, and can force the user to think in particular paradigms. This is not a new problem. The Metric System was created in 1795 to end such confusion; however, the babble of units continued, becoming more subtle as science advanced in the 19th Century. We were not so much arguing over land being measured in square metres or 'labouring days' (that is, the surface of land that a single man with his horse and plough could till in one working day), but the strength of a magnetic field. In 1899, the German physicist Max Planck proposed a new system of units, designed to be particularly useful in the newly invented areas of atomic physics and cosmology. Planck units are a set of measurement parameters defined exclusively in terms of four constants of Nature, in such a manner that these physical constants take on the numerical value of 1 when expressed in terms of these units. The Metric System was based on a measure of length (the metre), which was defined as one ten millionth of the circumference of a Quadrant of the Earth. So, the original Metric System was purely an Earth-based measurement system. Today, the metre is defined by the speed of light, and not a survey of the Earth's surface, and so is not restricted to this planet. The four constants that, by definition, have a numeric value, 1, when expressed in Planck units are[2]:

- the speed of light in a vacuum, c,
- the gravitational constant, G,
- the reduced Planck constant, $\hbar$,
- the Boltzmann constant, k_B.

Max Planck realized that if we rescaled the constants, which today define the SI units of metre, kilogram and second, these constants would no longer appear in physical equations—they would effectively be rescaled to unity. Noting that the dimensions of these physical constants are $[c] = \mathsf{LT^{-1}}$, $[G] = \mathsf{L^3\,M^{-1}T^{-2}}$, $[\hbar] = \mathsf{ML^2T^{-1}}$ and $[k_B] = \mathsf{ML^2T^{-2}}\,\theta^{-1}$, Planck determined length, mass, time and temperature scales:

[2] Planck units do not incorporate an electromagnetic dimension. However, some authors choose to extend the system to electromagnetism by, for example, defining the electric constant ε_0 as having the numeric value 1 or $1/4\pi$ in this system.

Table 4.3. Some derived Planck quantities.

Derived unit (dimension)	Description	Approximate SI equivalent
Volume (L^3)	$l_p^3 = (\hbar G/c^3)^{3/2}$	4.22×10^{-105} m^3
Energy (ML^2T^{-2})	$E_p = m_p c^2 = \hbar/t_p$	1.95×10^9 J
Force (MLT^{-2})	$F_p = \hbar/t_p l_p = c^4/G$	1.21×10^{44} N
Density (ML^{-3})	$a_p = c/t_p$	5.15×10^{96} kg m^{-3}

l_p, m_p, t_p and T_p, which are called the Planck length, Planck mass, Planck time and Planck temperature, respectively:

$$
\begin{aligned}
l_p &= \sqrt{\frac{\hbar G}{c^3}} \approx 1.6 \times 10^{-35}\text{m}, \\
m_p &= \sqrt{\frac{\hbar c}{G}} \approx 2.2 \times 10^{-8}\text{kg}, \\
t_p &= \sqrt{\frac{\hbar G}{c^5}} \approx 5.4 \times 10^{-44}\text{s}, \text{ and} \\
T_p &= \sqrt{\frac{\hbar c^5}{G k_B^2}} \approx 1.42 \times 10^{32}\text{K}.
\end{aligned}
$$

These quantities can be readily derived using the product of powers method (the Rayleigh algorithm) of dimensional analysis. In this approach, one starts with the three physical constants $\hbar$, c and G, and attempts to combine them to derive a quantity whose dimension is, for example, mass (M). The formula sought is of the form, $m_p = (c)^a (G)^b (\hbar)^c$, where a, b and c are constants to be determined by equating the dimensions of both sides. Therefore, we may write (using the dimensions given above),

$$[c^a G^b \hbar^c] = M^1 L^0 T^0 = M^{-b+c} L^{a+3b+2c} T^{-a-2b-c}. \tag{4.11}$$

The solutions are: $a = ½$, $b = -½$ and $c = ½$. Thus, the Planck mass is: $m_p = c^{1/2}\ G^{-1/2}\ \hbar^{1/2} = \sqrt{(c\hbar/G)}$. As we will see later, dimensional analysis can only determine a formula up to a dimensionless multiplicative factor. There is no *a priori* reason for starting with the reduced Planck constant $\hbar$ instead of h, which differs from it by a factor of 2π. But by physical insight, we know that a factor of 2π must be there.

Planck time may be defined in an analogous manner, and like other definitions of time has units of seconds. We write:

$$t_p = (\hbar)^a (G)^b (c)^c, \text{ where } [\hbar^a G^b c^c] = M^0 L^0 T^1 = M^{a-b} L^{2a+3b+c} T^{-a-2b-c}. \tag{4.12}$$

This time, the solutions are: $a = b = ½$ and $c = -5/2$, and so, Planck time, $t_p = \hbar^{1/2}$ $G^{1/2}$ c−5 2 $= \sqrt{(\hbar G/c^5)}$.

As pointed out in chapter 1, systems of measurement units needed to describe physical quantities can be derived from a set of base units. Table 4.3 offers a few derived Planck units. As with the corresponding base units, their use is mostly confined to theoretical physics, because most of them are too large, or too small for conventional physics; in addition, there are large uncertainties in their values.

4.3.3 The fine-structure constant

The fine-structure constant, α, is another constant of Nature. But it is actually the ratio of two dimensionally-similar quantities; and is thus a quantity without dimension—it is a number with a value, of order, 1/137. It is the coupling constant, or measure of the strength of the electromagnetic force that governs how electrically-charged elementary-particles (e.g., electron, muon) and light (photons) interact. This quantity was introduced by A. Sommerfeld in 1916, to explain the observed splitting or fine-structure of the energy levels of the hydrogen atom. Sommerfeld extended the Bohr theory (see section 12.2) to include elliptical orbits and the relativistic dependence of mass on velocity. The quantity α, which is equal to the ratio $v_{n=1}/c$; where $v_{n=1}$ is the velocity of the electron in the first circular Bohr orbit and c is the speed of light in vacuum, appeared naturally in Sommerfeld's analysis and determined the size of the splitting or fine-structure of the hydrogenic spectral lines. Sommerfeld's theory had some early success in explaining experimental observations, but could not accommodate the discovery of electron spin. Today, we write the fine-structure constant as:

$$\alpha = (1/4\pi\varepsilon_0)(e^2/\hbar c) = \mu_0 c e^2/2h, \tag{4.13}$$

where e is the elementary charge, $\hbar = h/2\pi$ where h is the Planck constant, $\varepsilon_0 = 1/\mu_0 c^2$ is the electric constant (permittivity of vacuum) and μ_0 is the magnetic constant (permeability of vacuum). In the 2019 revised-SI, c, and ε_0 are exactly known constants. The dimensions of these variables are given in table 4.4.

Let us see if we can use dimensional analysis to derive α. That is, if we assume

$$\alpha \propto (e)^a (h)^b (c)^c (\varepsilon_0)^d, \tag{4.14}$$

can we derive something that looks like equation (4.13)? The fine-structure constant is a dimensionless number; effectively, we are looking to find the simplest

Table 4.4. Variables for the fine-structure constant.

Variable	Dimension	Description
α		Fine-structure constant
e	IT	Elementary charge
h	$\mathrm{ML^2T^{-1}}$	Planck constant
c	$\mathrm{LT^{-1}}$	Speed of light
ε_0	$\mathrm{M^{-1}L^{-3}T^4I^2}$	Permittivity of space

dimensionless combination of e, c, h and ε_0. We may, therefore, write the dimensional equation as:

$$[\alpha] = (\mathrm{TI})^a(\mathrm{ML}^2\mathrm{T}^{-1})^b(\mathrm{LT}^{-1})^c(\mathrm{M}^{-1}\mathrm{L}^{-3}\mathrm{T}^4\mathrm{I}^2)^d = \mathrm{M}^{b-d}\mathrm{L}^{2b+c-3d}\mathrm{T}^{a-b-c+4d}\mathrm{I}^{a+2d}. \quad (4.15)$$

Note the coupling of e and ε_0 in the equation defining the contribution of the dimension of charge to the problem; this tells us immediately that we are dealing with an electromagnetic force. The solutions are readily found: $a = 2$, $b = d = c = -1$. We may therefore write,

$$\alpha = e^2/h.\ c.\ \varepsilon_0, \quad (4.16)$$

which, allowing for numerical constants, is the same as equation (4.13). However, the numerical constants that are rendered invisible by dimensional analysis, may be inferred from a knowledge of physics; that is, ε_0 always occurs as $4\pi\varepsilon_0$, and h occurs as $\hbar$.

4.3.4 Drop spatter analysis

Let us now look at drops impacting onto a surface; that is, spatter patterns. We imagine a drop of liquid falling from a source, and then encountering a solid surface. This collision, or impact, often results in a distinct spatter pattern—see figure 4.2. Examination of the patterns left by the drops after they have hit the surface often reveals spines, or tendrils spreading out from the central, roughly circular area. That is, the drop underwent some dynamical process—smaller droplets being liberated from the main drop as a consequence of its impact. Figure 4.2 displays such a pattern of spines radiating from the central flattened drop. The analysis of such images is one of the bases of forensic science, and the examination of crime scenes [3–5].

First, we must examine what happens when a drop of fluid falls onto a surface. The drops themselves are due to an instability in the fluid dynamics associated primarily with a competition between inertial forces and surface tension forces—this is why a continuous stream of fluid breaks up into drops (see analysis below). So, we can say immediately that the pattern of drop impacts will be insensitive to the viscosity of the liquid. Hence, for our dimensional analysis of this phenomenon; that is, an examination of the origin and number of the spines radiating from the large central spot (see figure 4.2(b)), will be based on those physical variables that are likely to be the most important. These variables are listed in table 4.5 (for the moment, ignore the bottom line in bold in table 4.5).

The spines are the result of tiny drops emerging from the main drop (see figure 4.2(a)). There is some geometric similarity here after impact, but the pattern of the spines is not strictly similar. But all the drops will be similar before impact. To rationalise what is happening upon impact, we hypothesise the function:

$$N = f(v, \rho, \sigma, r). \quad (4.17)$$

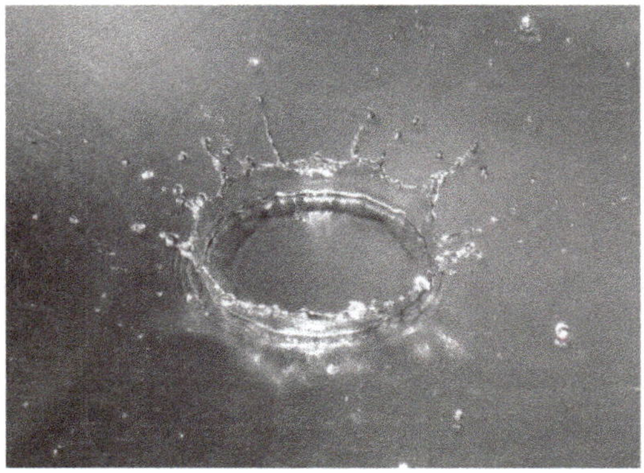
(a)

(b)

Figure 4.2. (a) Origin of spatter pattern. The smaller droplets flying out along spines from the spatter of a large drop will yield a series of smaller satellite drops, and continuous spines radiating from the main drop—as seen in (b), which is a typical spatter pattern. Image (a) This 'A drop of water splashing onto a hard surface' image has been obtained by the author from the Wikimedia website where it was made available under a CC BY-SA 3.0 licence. It is included within this article on that basis. Image(b) This 'blood splot' has been obtained by the author from the Wikimedia website where it was made available under a CC BY-SA 3.0 licence. It is included within this article on that basis. It is attributed to Karta24.

Table 4.5. Variables for spatter pattern.

Variable	Dimensions	Description
N		Number of spines radiating from a splattered drop. This is a dimensionless number.
v	LT^{-1}	Velocity of drop on impact.
ρ	ML^{-3}	Density of fluid
σ	MT^{-2}	Surface tension
r	L	Radius of drop on impact
μ	$ML^{-1}T^{-1}$	**Dynamic viscosity of the fluid**

Note that N is dimensionless, but is not a constant; the variables are given in table 4.5. A straightforward application of dimensional analysis leads to the result (see equation (3.14)),

$$N = f(\rho v^2 r/\sigma). \tag{4.18}$$

This is another of the standard dimensionless numbers, which appear in fluid mechanics. This is a Weber number (We), named for the German engineer Moritz Weber (1871–1951). The Weber number relates inertia and surface tension as, $\mathrm{We} = \rho v^2 r/\sigma = \rho v^2 r^2/\sigma r = \frac{\text{Inertial forces}}{\text{Surface tension forces}}$. Inertial forces will enhance the ability of the fluid to spread over a surface, thus encouraging the appearance of satellite features and spines. Surface tension forces seek to keep the drop intact, even after it has hit the surface—the surface tension is responsible for the curvature of drops on a flat surface. When a liquid drop lands on a solid surface without wetting it, it is seen to bounce with some elasticity. The manner in which a water-drop, of radius, r, deforms during its impact with a highly hydrophobic solid (plastic surface) depends mainly on its impinging velocity, v. The Weber number compares the kinetic and surface energies of the drop, where ρ and σ are the liquid density and surface tension, respectively. The greater the value of We, the larger are the deformations that occur during the impact.

High-speed photography is the means of studying the impact of drops on surfaces (with contact time, τ), and forensic science has developed technology to probe the origins of certain types of spattered drops; particularly, blood-spatter patterns [3–5]. As the impact is mainly inertial, τ is expected to be a function of only r, v, ρ and σ, and thus to vary as $r/v\ f(\mathrm{We})$. For a Hertz shock response, for example, the maximum vertical deformation, d, scales as: $r(\rho^2 v^4/E^2)^{1/5}$, where E is the Young's modulus of the drop. Taking a drop's Laplace pressure, $E \approx \sigma/r$, as an equivalent modulus and noting that $\tau \approx d/v$, we find for a Hertz drop that $f(\mathrm{We}) \sim \mathrm{We}^{2/5}$ and that the contact time varies as $v^{-1/5}$ and $r^{7/5}$ [6]. The scaling for τ is the same as for the period of vibration of a drop derived by Lord Rayleigh [7].

Interestingly, the brevity of the contact means that a drop containing surfactants, which will spread when gently deposited onto the solid, can bounce when dropped onto it; this is because the contact time is too short to allow the adsorption of the surfactants onto the fresh interface generated by the shock. Conversely, the contact time should provide a measurement of the dynamic surface tension of the drop.

Establishing that the analysis of the variables responsible for the spines around an impacted drop of a fluid such as blood is a Weber number, is as far as dimensional analysis can take us with this phenomenon. To go further, we require experimental input. Such a study may be found in [3]. Here, Adam investigated the number of spines generated upon impact (N) with the Weber number; that is, the velocity and diameter of the drop—the density and the surface tension of the fluid would have been constants, as they are properties of the fluid. Adam found that when the number of spines was plotted against the square root of the Weber number, he obtained a straight line passing through the origin. However, above $\sqrt{\mathrm{We}} \sim 30$, there was a change to a non-linear regime. That is, a scaling law $N \propto \mathrm{We}^{1/2}$ was found to well describe (collapse) the experimental data below a certain limit.

Given that experiment tells us something changes in the dynamics of this phenomenon when the Weber number, We, goes above 900, we need to go back to our initial assumptions. In this region of the Weber number there is a range of values of N for each value of We. This could mean an addition variable, because the density and the surface tension are constants, and so we are only seeing new dynamics as the drop size and the terminal velocity of the drop increases. So, we will reconsider the dimensional analysis of this problem, but after adding an additional variable—the possible effect of viscosity (the bottom line in bold in table 4.5). This idea is strengthened by the arguments of Attinger *et al* [4], who maintain that in experiments on impacting drops at high Weber number, viscosity should be included. Certainly, at higher Weber number, the simple two variable solution of the dimensional analysis $N = F(\rho v^2 r/\sigma)$ is insufficient to collapse all the experimental data. Our new hypothesis is,

$$N = f(v, \rho, \sigma, r, \mu). \tag{4.19}$$

Applying the Buckingham Π-theorem (which we shall look at in the next chapter) to all the variables in table 4.5; there are six variables (N is a number) and three base dimensions (the mechanical dimensions), so there will be three dimensionless Π-groups. One of these three groups will be the dimensionless N, another will be the Weber number, and the third can be inverted to obtain a Reynolds number (Re), thus,

$$N = f_2\left(\rho v^2 r/\sigma, \frac{\mu}{\rho v r}\right) \rightarrow G\left(\rho v^2 r/\sigma, \frac{\rho v r}{\mu}\right) = G(\mathrm{We}, \mathrm{Re}). \tag{4.20}$$

We see that including viscosity has increased the complexity of the hypothesised function of nondimensional parameters from two variables to three. We would now need to perform a new set of experiments to see if we could collapse all the measurements of N at a variety of values of v and r; that is, a single value of N on a surface defined by Re and We.

References

[1] White F 2011 *Fluid Mechanics* 7th edn (New York: McGraw Hill) ch 5

[2] Taylor G I 1950 The formation of a blast wave by a very intense explosion (I): theoretical discussion Proc. of the Royal Society of London. Series A, Mathematical and Physical Sciences **201** 159–74

Taylor G I 1950 The formation of a blast wave by a very intense explosion (II): the atomic explosion of 1945 Proc. of the Royal Society of London. Series A, Mathematical and Physical Sciences **201** 175–86

[3] Adam C D 2012 Fundamental studies of bloodstain formation and characteristics *Forensic Sci. Int.* **219** 76–87

[4] Attinger D, Moore C B, Donaldson A, Jafari A and Stone H A 2013 Fluid dynamics topics in bloodstain pattern analysis: comparative review and research opportunities *Forensic Sci. Int.* **231** 375–96
[5] Knock C and Davison M 2007 Predicting the position of the source of blood stains for angled impacts *J. Forensic Sci.* **52** 1044–49
[6] Richard D, Clanet C and Quéré D 2002 Contact time of a bouncing drop *Nature* **417** 811
[7] Rayleigh L 1879 VI: on the capillary phenomenon of jets *Proc. R. Soc. Lond.* A **29** 71–97

Chapter 5

The Buckingham Π-theorem and its application

'Some proofs command assent. Others woo and charm the intellect. They evoke delight and an overpowering desire to say, 'Amen, Amen''.

Lord Rayleigh

Dimensional analysis is a method for reducing complex physical problems to their simplest (most economical) forms, prior to quantitative analysis or experimental investigation [1, 2]. Its use in science and engineering is ubiquitous. Applications are many; including, astrophysics, electromagnetic theory, radiation, aerodynamics, ship design, heat and mass transfer, mechanics of structures, explosions, chemical reactions and processing [3–5], biology [6, 7] and even economics [8]. As we have seen in the examples given in chapter 4, dimensional analysis reduces a problem's degrees of freedom to the minimum, and thus suggests the most economical scaling laws. It can be particularly useful in exploratory investigations of novel phenomena for which the equations and boundary conditions have not yet been fully delineated.

One of the central ideas behind dimensional analysis is the generation of a relation, an equation that links together the variables believed to be important in the phenomenon under investigation. How are the variables related, and what is the dependence of a particular variable on one, or more other variables? We saw something of this in the analysis of drop formation in section 4.3.4. Lord Rayleigh introduced a systematic method for investigating the relationship between the base dimensions of a set of variables in *The Theory of Sound* of 1877. In this chapter we will look at the Buckingham Π-theorem, which attempts to systematise the algorithm of Lord Rayleigh to generate a formalism for generating a quantity equation; a formalism which has been widely adopted.

According to the International Vocabulary of Metrology [9], a mathematical relation between n quantities that is independent of measurement units is called a quantity equation. The theorem, due to the American metrologist Edgar

doi:10.1088/978-0-7503-3655-0ch5

Buckingham (1867–1940) [10], states that such a quantity equation can be written in the form $f(\Pi_1, \Pi_2, \ldots, \Pi_k) = 0$, where the Π_is are dimensionless combinations of the original quantities or variables, and k is the difference between the number of base dimensions observed in the n variables (m). In table 4.2, we looked at the variables used by Taylor to calculate the yield of a nuclear explosion. This table contained four variables ($n = 4$), composed of the three mechanical base dimensions (M, L and T), so $m = 3$, and the four variables could therefore be written as, $k = 4–3$, or a single nondimensional Π-group. It is often the case in a new project that one knows the quantities, q_i, that are most relevant for describing a particular phenomenon, but not the form of the corresponding quantity equation. As it is usually the case that n is greater than m, it is easier to find from experiment the values of the Π_is satisfying $f(\Pi_1, \Pi_2, \ldots, \Pi_k) = 0$ than to find the values of the q_is satisfying $f(q_1, q_2, \ldots, q_m) = 0$. Once the former task has been accomplished, the values of the q_is can be obtained from the Π_is. [11].

The Buckingham Π-theorem: Let $q_1, q_2, q_3 \ldots q_n$ be n dimensional variables that are relevant in a given problem, and are related by an (unknown) dimensionally homogeneous set of equations. These variables be expressed via a functional relationship of the form, $F(q_1, q_2, q_3 \ldots q_n) = 0$, or equivalently $q_1 = f(q_2, q_3 \ldots q_n)$.

If m is the number of base dimensions required to describe the n variables, then there will be m primary variables, and the remaining $k = (n-m)$ variables can be expressed as dimensionless and independent quantities, or Π-groups, $\Pi_1, \Pi_2, \ldots \Pi_{n-m}$. The functional relationship sought can thus be reduced to the much more compact form: $\varphi(\Pi_1, \Pi_2, \ldots \Pi_{n-m}) = 0$, or equivalently $\Pi_1 = \varphi(\Pi_2, \ldots \Pi_{n-m})$.

Most textbooks that cover dimensional analysis use Buckingham's theorem as the starting point. But it is not the theorem that is of interest to the scientist. It is that: (*i*) any quantity equation $f(q_1, q_2, \ldots, q_n) = 0$ can be put in an equivalent dimensionless form $f(\Pi_1, \Pi_2, \ldots, \Pi_m) = 0$, where m is less than n; (*ii*) that if the latter quantity equation is unknown it can be obtained empirically—the difficulty increasing with increasing n; and (*iii*) that for the method to yield useful results all relevant quantities q_i should be listed as arguments of the original unknown function.

It should also be stressed that the dimensionless quantities Π_is are not unique (as we will see below), and that the method for obtaining them is, to a large extent, arbitrary. Indeed, in simple cases they can be found by inspection; otherwise, there are two procedures for their derivation. The first is the one expounded in most textbooks, which consists in setting up an underdetermined system of linear algebraic equations for the exponents to which the quantities q_i should be raised to powers, such that their product is dimensionless (this is an application of the Rayleigh algorithm). This procedure is actually not very convenient, as such a system has, in general, to be solved several times; each time making arbitrary choices for some of the exponents. In addition, one should apply hindsight, or experience for such choices so as to avoid 'inconvenient' dimensionless numbers. The second procedure is more intuitive. It consists in sequentially eliminating all base dimensions, one at a time and in any order, by arbitrarily choosing one of the quantities that contains the dimension to be eliminated. So, if the original quantity equation consists of n quantities having m base dimensions, eliminating the latter would normally lead to $k = n - m$ dimensionless quantities. Buckingham's theorem then follows.

There exist numerous instances in which some of the independent variables that define the variables, q_i, have essentially invariant values in all of the cases under consideration (they may be numerical, or dimensional constants; e.g. angles and numerical constants). The question then arises: does this lead to a further simplification; that is, to an additional reduction in the problem's inherent degrees of freedom? If so, how is the process of dimensional analysis, and with it the Π-theorem, altered? Simply omitting the quantities that have fixed values and performing dimensional analysis on the remainder will not answer this question. Dimensional analysis must be based on a complete set of independent variables that define the phenomenon of interest; that is, all quantities with values that may affect the quantity of interest must be included regardless of whether some have invariant values in the cases that are under consideration. Omitting even one relevant independent variable can ruin the analysis. Perhaps Lord Rayleigh exaggerated a bit when he wrote (see Preface) '*It happens not infrequently that results in the form of 'laws' are put forward as novelties on the basis of elaborate experiments, which might have been predicted a priori after a few minutes consideration.*' While we might take longer to undertake a dimensional analysis, Rayleigh's authoritative statement still holds true: dimensional analysis is indeed a useful tool for providing an insight into physical laws, and how they shape the world around us. We will apply the Buckingham Π-theorem to a number of different systems. The details of the proof of the theorem are not relevant, and may be found in the appropriate references [1–5, 10 (1914 reference)].

5.1 The Buckingham Π-technique

The Buckingham Π-technique can be thought of as a formal procedure for deriving a quantity equation for a list of variables. There are several steps, which are outlined below (also see chapter 6).

- Define the problem, and list the essential variables. Identify which is the main variable of interest (the dependent variable); that is, $q_1 = f(q_2, q_3 \ldots q_n)$. It is important to think physically about the problem (see chapter 3); are there any constraints; for example, the weight of an object is a force ($\mathsf{MLT^{-2}}$); so, $F_w = \rho g l^3$ (here ρ is the density, $\mathsf{ML^{-3}}$, g is the acceleration due to gravity ($\mathsf{LT^{-2}}$) and l is a length) of the term on the RHS, only two components are independent—unless g is also a variable.
- Express each of the n variable in terms of its base dimensions.
- Determine the number of Π-groups; there will be $n - m = k$ such groups, where m is the number of base dimensions. Then select primary, and the repeating variables (j). The primary variables tend to be the variables of interest, and

the repeat variables will contain the base dimensions of number m. For example, if mass, length and time each appear in, at least, one variable, j is set to 3. As the Buckingham method progresses, it often becomes evident that there is a problem. In such cases, j should be reduced by 1 and the analysis repeated. Once j is found, the number of dimensionless parameters (or Π-groups) expected is $k = n-j$, where k is the number of Π-groups. This equation relating k to n and j is at the heart of the Buckingham Π-Theorem.

- Generate the dimensionless Π-groups, and check they are all dimensionless. The Π-groups are formulated by multiplying each of the remaining variables (those that were not chosen as repeating variables) in turn by the repeating variables, each in turn raised to some unknown exponent. The exponents are found algebraically by forcing the Π-group to be dimensionless. The convention is to form the first Π-group using the dependent variable. Note that Π-groups can be adjusted after they are formed in order to agree with the dimensionless groups commonly used in the literature; e.g., Re, We, etc (see discussion of section 4.2).
- Express the result in the form, $\Pi_1 = \varphi(\Pi_2, \ldots \Pi_{n-m})$, where Π_1 contains the quantity of interest, and interpret your result physically.
- Lastly, but not least: Compare with experimental data.

5.2 Some examples of dimensional analysis involving the Π-theorem

In the following sections one can see some examples of dimensional analysis involving the Buckingham Π-theorem.

5.2.1 Dimensional analysis of a soap bubble

It is known that the pressure inside a bubble (see figure 5.1) must be greater than that outside, and that surface tension acts like a skin to preserve this pressure difference (Δp = pressure difference across the bubble boundary). The pressure difference, Δp, is thus a function of surface tension (σ), and bubble radius (r). Let us examine this problem with the Buckingham Π-technique, following the steps outlined above. The number of variables in the problem is 3; $n = 3$ (and these are listed in table 5.1). So, we may write our initial hypothesis of a relationship between the pressure difference inside and outside of the bubble, and the surface tension of the bubble as:

$$\Delta p \propto (\text{surface tension, } \sigma)(\text{radius of bubble, } r). \tag{5.1}$$

From table 5.1, mass, length and time are the only base dimensions represented by this set of variables. Thus, $m = 3$. This yields $k = n-m = 0$; that is, we expect zero Π-groups from the dimensional analysis. When this situation arises, either we do not have enough variables in the original hypothesis (equation (5.1))—that is, not enough physics is represented by the incomplete list of variables (that is, n is too small), or m is incorrect. If there are no additional variables, to proceed we reduce m by 1, which yields $k = n-m = 1$; that is, we expect a single Π-group from the dimensional analysis.

Figure 5.1. A soap bubble. This 'a soap bubble' has been obtained by the author from the Wikimedia website where it was made available under a CC BY-SA 3.0 licence. It is included within this article on that basis. It is attributed to Brocken Inaglory.

Table 5.1. Variables for the properties of a soap bubble.

Variable	Dimension	Description
Δp	$\mathrm{ML^{-1}\,T^{-2}}$	Pressure difference
σ	$\mathrm{MT^{-2}}$	Surface tension
r	L	Bubble radius

We need to pick two repeating variables; the pressure difference is not a good choice as it is the dependent variable. The only other choices are surface tension and bubble radius. There is only one Π-group, and it is found by combining the remaining (dependent) variable with the two repeating variables to form a Π-group, as follows: $\Pi = \Delta p\ (\sigma)^a\ (r)^b$. To find the exponents, we force the group to be dimensionless. Thus, the dimensional equation is

$$\mathrm{M^0L^0T^0} = \mathrm{ML^{-1}T^{-2}(MT^{-2})}^a\mathrm{(L)}^b = \mathrm{M}^{a+1}\mathrm{L}^{b-1}\mathrm{T}^{-2a-2}. \tag{5.2}$$

The results are: $a = -1$, and $b = 1$, so, $\Pi = \Delta p\ r/\sigma$, or,

$$\Delta p\ \ r/\sigma = \text{constant}, \quad \text{or } \Delta p = \text{constant}.\ \sigma/r. \tag{5.3}$$

Notice that instead of a dependent variable as a function of two independent variables, the problem has been reduced to one dependent parameter as a constant. In cases such as this where there is only a single Π-group, that Π-group must be a constant. This is an example of the power of dimensional analysis. Here we have obtained a functional relationship between the pressure, radius and surface tension of a bubble to within a constant of proportionality with minimal physical insight. Dimensional analysis cannot provide the numerical constant, but it can provide information about how one variable depends upon others. Exact analysis provides the constant of proportionality in equation (5.3); namely, a numerical factor, 4.

We will now consider the bubble in greater detail. Observation permits us to see that such bubbles are subject to wind and air currents, and move with a velocity, v; so they have a buoyancy. When the air current disappears, they will fall due to gravity and be subject to a drag force (F_d) in the air—they will have a terminal velocity. Consequently, there will be three forces operating on a large bubble in the real world: the downward force due to gravity, $(m_b+m_a)g$, where m_a is the mass of the air inside the bubble, m_b is the mass of the fluid that forms the bubble, and g is the acceleration due to gravity; the upward buoyancy force, $gV\rho$, where V is the volume inside the outer surface of the bubble and ρ is the density of the surrounding air; and the drag force, F_d that opposes the gravity force. We will see later that this drag force depends on the surface area, A of the bubble, and on the viscosity of the air, μ (because of friction between the air and the bubble) and also on the air density (because the surrounding air is set into motion, so its inertia is involved). The preliminary quantity equation then becomes: $f(V, m_b, m_a, g, v, \rho, A, \mu) = 0$, where we have eight variables (compared to the three above in our first look at the bubble). However, not all these variables are functionally independent. Both the volume and the surface area depend on the diameter, D, of the bubble. Moreover, if the temperature and pressure of the air inside the bubble are not too different from those of the outside, m_a can be taken as proportional to the product of ρ and D^3. In this way, we arrive at:

$$f(v, m_b, g, D, \rho, \mu) = 0.$$

We are now down to six variables (n). Invoking the Buckingham theorem, these six variables are defined by three base dimensions (m); consequently, there are three dimensionless Π-groups to be identified (k). To identify these three dimensionless groups of variables, we will first use a process of elimination—of multiplication and division of the variables, so as to cancel all the base dimensions. (The dimensions involved are given in table 5.2, with the addition of $[m_b]$ = M, and $[g]$ = LT^{-2}.) We arbitrarily select one of the three base dimensions, L, M or T, and also select any one of the variables that contain that dimension in order to eliminate it from the other variables; for example, eliminating a dependence on length by dividing and multiplying by D gives:

$$f(vD^{-1}, m_b, gD^{-1}, D^3\rho, D\mu) = 0.$$

One variable, D, has been lost as it is now dimensionless. Continue the elimination process by selecting M, via m_b, which gives:

$$f(vD^{-1}, gD^{-1}, m_b^{-1}D^3\rho, m_b^{-1}D\mu) = 0,$$

and finally eliminate T via $m_b^{-1}D\mu$. We then arrive at: $f(\Pi_1, \Pi_2, \Pi_3) = 0$, where $\Pi_1 = vm_b/D^2\mu$, $\Pi_2 = m_b^2g/D^3\mu^2$ and $\Pi_3 = D^3\rho/m_b$.

As mentioned above, the Π-theorem technique does not generate a single answer—the nature of the dimensionless Π-groups precludes a unique result. By choosing other elimination paths, other dimensionless quantity equations are obtained; for example, $f(\Pi_1', \Pi_2', \Pi_3') = 0$, where $\Pi_1' = \Pi_1\Pi_3 = vD\rho/\mu$, $\Pi_2' = \Pi_2\Pi_3 = m_bg\rho/\mu^2$ and

$\Pi_3' = \Pi_1\Pi_3^2 = gD^3\rho^2/\mu^2$. In fact, since the product of dimensionless quantities raised to any exponents is also dimensionless, a limitless variety of equivalent quantity equations having three dimensionless quantities as arguments is possible. For example, $(\Pi_1)^{\ln 2}(\Pi_2)^{1/\pi}$ would be a perfectly legitimate (though unusual) dimensionless quantity. Note that Π_1' is the Reynolds number, $\mathrm{Re} \equiv vD\rho/\mu$.

Let us now apply the Buckingham Π-theorem to the same set of variables, and compare the results of a quantity equation devised using repeat variables. We have, $f(v, m_b, g, D, \rho, \mu) = 0$; we seek to construct three dimensionless Π-groups from the three lead variables (k=3): v, m_b and μ, and the j=3 repeat variables: ρ, g and D. The dimensionality of the first Π-group is:

$$\begin{aligned}\Pi_1 &= \mu(\rho)^a(g)^b(D)^c = \mathsf{ML^{-1}T^{-1}}(\mathsf{ML^{-3}})^a(\mathsf{LT^{-2}})^b(\mathsf{L})^c \\ &= \mathsf{M}^{a+1}\mathsf{L}^{-3a+b+c-1}\mathsf{T}^{-2b-1}.\end{aligned} \tag{5.4}$$

The solutions are: $a = -1$, $b = -½$ and $c = -3/2$. So, $\Pi_1 = \mu/\rho\sqrt{g}\sqrt{D^3} = \mu^2/\rho^2 gD^3$. The second and third Π-groups are found analogously.

$$\begin{aligned}\Pi_2 &= m_b(\rho)^a(g)^b(D)^c = M(\mathsf{ML^{-3}})^a(\mathsf{LT^{-2}})^b(\mathsf{L})^c \\ &= \mathsf{M}^{a+1}\mathsf{L}^{-3a+b+c}\mathsf{T}^{-2b}.\end{aligned} \tag{5.5}$$

The solutions are: $a = -1$, $b = 0$ and $c = 0$. So, $\Pi_2 = m_b/\rho D^3$.

$$\begin{aligned}\Pi_3 &= v(\rho)^a(g)^b(D)^c = \mathsf{LT^{-1}}(\mathsf{ML^{-3}})^a(\mathsf{LT^{-2}})^b(\mathsf{L})^c \\ &= \mathsf{M}^{a}\mathsf{L}^{-3a+b+c+1}\mathsf{T}^{-2b-1}.\end{aligned} \tag{5.6}$$

The solutions are: $a = 0$, $b = -½$ and $c = -½$. So, $\Pi_3 = v/\sqrt{g}\sqrt{D} = v^2/gD$. This third dimensionless group is the square of the well-known Froude number (Fr), $\mathrm{Fr} \equiv v/\sqrt{gl}$, defined as the ratio of the flow inertia to the external field—the latter in many applications simply due to gravity. Using $f(\Pi_1, \Pi_2, \Pi_3) = 0$, and $\Pi_1 = f(\Pi_2, \Pi_3)$ we may write for the Buckingham Π-analysis of the bubble system:

$$\mu^2/\rho^2 gD^3 = f(m_b/\rho D^3, v^2/gD) = f(m_b/\rho D^3, \mathrm{Fr}^2). \tag{5.7}$$

The equivalent expression for the elimination solution of the same system (the same variables), given above is:

$$vm_b/D^2\mu = f(m_b^2 g/D^3\mu^2, D^3\rho/m_b). \tag{5.8}$$

Of course, the labelling of the Π-groups depends upon the manner of the solution. But it is clear that there are differences in the results of the two analyses; the variable m_b (of dimension M) occurs in each of three dimensionless groups from the elimination method, but only in one of the three dimensionless groups from the Buckingham analysis. Likewise, g appears in two of the dimensionless groups from the Buckingham analysis, but only in one group from the elimination analysis. We see that the Π_2 of the Buckingham solution is the inverse of Π_3 of the elimination solution; given the properties of Π-groups, this means they are equivalent. In addition, using the properties of the dimensionless Π-groups one can bring the results of the two methods of dimensional analysis closer. The Froude number appeared directly in the Buckingham technique; however, if we let $\Pi'' = \Pi_1^2 \Pi_2^{-1}$, where Π_1 and Π_2 are as defined above in the elimination analysis; then $\Pi'' = v^2/Dg$, the Froude number. In addition, the Π_1 of the Buckingham analysis can be formed from the Π-groups of the elimination analysis. The product, $\Pi_2\Pi_3^2$ of the elimination analysis is $gD^3\rho^2/\mu^2$, which is the inverse of Π_1 derived from the Buckingham analysis. So the two methods do generate the same results, but additional analysis is required[1].

This analysis demonstrates that there is no set of unique solutions to a problem in dimensional analysis. Not only does dimensional analysis lose all the numerical factors in a problem, but depending upon the manner of the solution one is able to generate a number of possible solutions. The more complex the problem; that is, the greater the number of variables involved, the greater the number of possible, equivalent combinations of those variables that are generated.

Deriving an analytical form for a quantity equation such as $f(\Pi_1, \Pi_2, \Pi_3) = 0$ may be difficult, or even impossible. However, in principle, a model could be tested empirically by measuring some of the original quantities as the others are varied. In the bubble example, one could measure the terminal speeds of bubbles having different volumes, under different ambient conditions. These data would allow us to construct a surface satisfying $\Pi_1 = f(\Pi_2, \Pi_3)$. Naturally, the difficulty of the experiment increases geometrically as the number of dimensionless quantities increases, so one should seek to identify only the most relevant variables that influence the phenomenon under investigation.

5.2.2 Lift generated by a wing

Consider the case of incompressible flow over an airplane wing (figure 5.2). Wing lift (L) is known to depend on the flow speed (v) of the air over the wing, angle of attack (α), chord length (c) of the wing, the density (ρ) and viscosity (μ) of the fluid support (air) [7]; these variables are defined in table 5.2. Let us examine this problem with the Buckingham Π-technique. Here, the number of variables is six. As we wish to characterise the lift force, we may write for our hypothesised model:

$$L \propto F(v, \alpha, c, \rho, \mu). \tag{5.9}$$

Table 5.2 tells us that we have six variables, and that the five dimensioned variables are composed of three base dimensions. Thus, m=3, which yields $k = n - m = 3$; that is, we expect three Π-groups from the dimensional analysis.

Choose j repeating variables: Here we need to pick three repeating variables. Lift force is not a good choice since it is the dependent variable. Angle of attack is not permitted as it is already dimensionless. (Note that angle of attack will be a dimensionless Π-group by itself, Π_3.) Out of the remaining four variables, viscosity is the least basic or desirable variable to be repeated in all the Π-groups (it contains all

[1] There are a lot of deep rabbit holes in dimensional analysis.

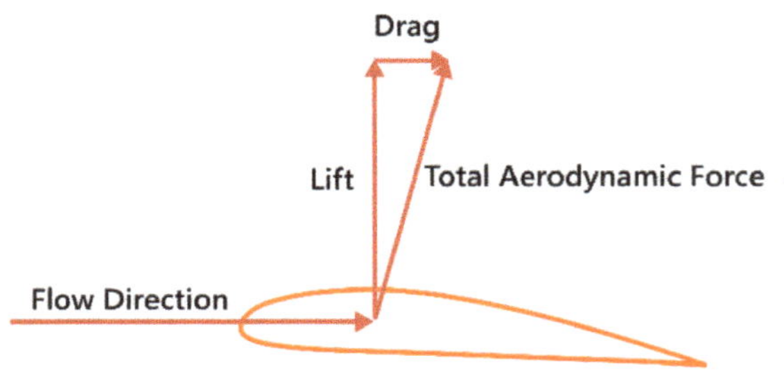

Figure 5.2. Lift is defined as the component of the aerodynamic force that is perpendicular to the flow direction, and drag is the component that is parallel to the flow direction. This 'airfoil lift and drag' image has been obtained by the author from the Wikimedia website where it was made available under a CC BY-SA 4.0 licence. It is included within this article on that basis. It is attributed to J Doug McLean.

Table 5.2. Variables for problem of lift.

Variable	Dimension	Description
L	MLT^{-2}	Lift force
v	LT^{-1}	Velocity of air
c	L	Chord length
ρ	ML^{-3}	Density of air
μ	$ML^{-1}\ T^{-1}$	Viscosity of air
α	(dimensionless)	Angle of incidence, or attack

three base dimensions). The best variable repeat variables are: density (ρ), velocity (v) and chord length (c). We will first generate a dimensionless group (Π_1) with the lift force, since it is the dependent variable. Thus,

$$\Pi_1 = L(v)^a(c)^b(\rho)^c. \tag{5.10}$$

The dimensional equation is

$$M^0L^0T^0 = MLT^{-2}(LT^{-1})^a(L)^b(ML^{-3})^c = M^{c+1}L^{a+b-3c+1}T^{-a-2}. \tag{5.11}$$

The solutions are: $a = -2$, $b = -2$ and $c = -1$; thus,

$$\Pi_1 = L/v^2c^2\rho. \tag{5.12}$$

Likewise, we construct the second Π-group using viscosity, and the repeating variables:

$$\Pi_2 = \mu(v)^e(c)^f(\rho)^g. \tag{5.13}$$

The new dimensional equation is

$$M^0L^0T^0 = ML^{-1}T^{-1}(LT^{-1})^e(L)^f(ML^{-3})^g = M^{g+1}L^{e+f-3g-1}T^{-e-1}. \tag{5.14}$$

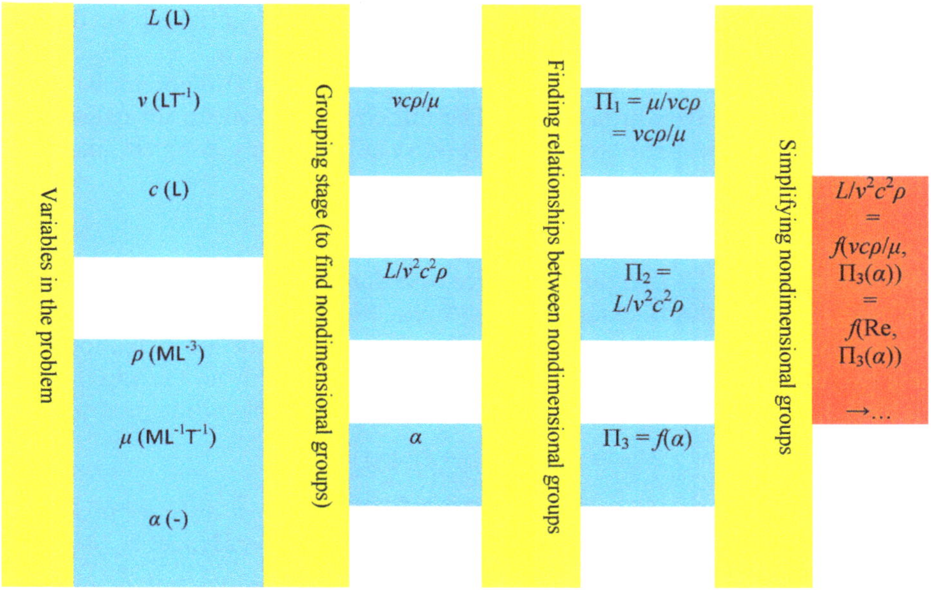

Figure 5.3. Outline for the dimensional analysis of lift force. After examining the problem, we identify six relevant variables (see table 5.2). The variables go into the grouping stage, which relates the base dimensions (three) in the variables (six), and generates three dimensionless Π-groups. The relation finder produces the most general relation using the three groups. The simplifier transforms the relation into a form more convenient for collapsing experimental data. In this case, however, further experiment (or insight) is required to fully define the problem.

The solutions are: $g = -1$, $e = -1$ and $f = -1$. Thus,

$$\Pi_2 = \mu/vc\rho, \quad \text{or } \Pi_2 = vc\rho/\mu. \tag{5.15}$$

This Π-group has been inverted to match the Reynolds number. With the third Π-group being the angle of attack, the final functional relationship is:

$$L/v^2c^2\rho = f(vc\rho/\mu, \Pi_3(\alpha)) \rightarrow L/v^2c^2\rho = f(\text{Re}, \Pi_3(\alpha)). \tag{5.16}$$

Instead of a dependent variable as a function of five independent variables, the problem has been reduced to one dependent parameter as a function of only two independent parameters. The dependent Π-group on the LHS is a lift coefficient (which normally has a factor of 2 in for convenience), while the first independent parameter on the RHS is the Reynolds number. Figure 5.3 shows a summary of this analysis.

5.2.3 Flows in pipes, tubes and blood vessels

We will now look briefly at the modelling of fluid flow in conduits. First, we consider the case of a pressure change experienced by a fluid flowing through a fixed-diameter tube. The model may be applied to the study of blood pressure changes in selected arterial sections (fixed radius, and circular section), with an additional simplifying assumption that blood is a homogeneous mixture on the solution-like scale, and the flow velocity in the selected vessels is constant. With these assumptions, the pressure change, Δp, depends on four parameters: artery diameter, d; length of the investigated flow section, x; flow velocity, v; and blood viscosity, μ.

The variables: d, x, v, μ may be used in the Buckingham Π-theorem, with $n = 4$; the dimensions are given in table 5.3. The unknown we wish to investigate is the pressure drop, Δp, over the section of blood vessel being investigated, and we will hypothesis a relation:

$$\Delta p = f(d, x, v, \mu). \tag{5.17}$$

We see from table 5.3 that there are three base dimensions involved in this problem; thus, m =3, and so the Buckingham Π-theorem tell us there are two dimensionless groups of variables needed to describe our hypothesised function (equation (5.17)). These two dimensionless groups of variables are constructed from the variables given in table 5.3, using repeat variables. First, consider the variable of interest,

$$\Delta p = (d)^a(x)^b(v)^c(\mu)^d F(\Pi_1), \tag{5.18}$$

where Π_1 is a dimensionless variable. We chose: d, v and μ as a set of repeating variables representing the dimensions, L, T, M, respectively. Hence:

$$\Pi_1 = \Delta p(d)^{\mathrm{a}}(v)^{\mathrm{b}}(\mu)^{\mathrm{c}}. \tag{5.19}$$

This gives us a dimensional equation:

$$\begin{aligned}\mathsf{L}^0\mathsf{M}^0\mathsf{T}^0 &= \Delta p(\mathsf{L})^a(\mathsf{LT}^{-1})^b(\mathsf{ML}^{-1}\mathsf{T}^{-1})^c = \mathsf{ML}^{-1}\mathsf{T}^{-2}(\mathsf{L})^a(\mathsf{LT}^{-1})^b(\mathsf{ML}^{-1}\mathsf{T}^{-1})^c \\ &= \mathsf{M}^{c+1}\mathsf{L}^{a+b-c-1}\mathsf{T}^{-b-c-2}.\end{aligned} \tag{5.20}$$

With solutions: $a = 1$, $b = -1$ and $c = -1$. So, $\Pi_1 = \Delta p\ d/v\mu$. The second nondimensional group, Π_2, is found analogously using the other variable not included in the set of repeat variables:

$$\Pi_2 = x(d)^{\mathrm{a}}(v)^{\mathrm{b}}(\mu)^{\mathrm{c}}. \tag{5.21}$$

This gives us a second dimensional equation,

$$\begin{aligned}\mathsf{L}^0\mathsf{M}^0\mathsf{T}^0 &= x(\mathsf{L})^a(\mathsf{LT}^{-1})^b(\mathsf{ML}^{-1}\mathsf{T}^{-1})^c = \mathsf{L}(\mathsf{L})^a(\mathsf{LT}^{-1})^b(\mathsf{ML}^{-1}\mathsf{T}^{-1})^c \\ &= \mathsf{M}^c\mathsf{L}^{a+b-c+1}\mathsf{T}^{-b-c}.\end{aligned} \tag{5.22}$$

Table 5.3. Variables for blood flow.

Variable	Dimension	Description
Δp	$\mathsf{ML}^{-1}\mathsf{T}^{-2}$	Pressure drop
d	L	Diameter of vessel
x	L	Length of vessel
v	LT^{-1}	Velocity of blood
μ	$\mathsf{ML}^{-1}\mathsf{T}^{-1}$	Viscosity of blood

Here, the solutions are: $b = c = 0$ and $a = -1$. So, $\Pi_2 = x/d$. We see that both Π-groups are dimensionless. The Π-theorem allows us to say $\Pi_1 = f(\Pi_2)$, and so:

$$\frac{\Delta p\, d}{v\mu} = f\left(\frac{x}{d}\right),$$

and

$$\Delta p = Cv\mu x/d^2, \tag{5.23}$$

where C is a constant of proportionality, which will contain all the numerical factors. Any statement with a greater degree of precision requires experimental data. We see how the pressure drop depends upon the geometry of the blood vessel; being inversely proportional to flow cross-section. The simple example shows that the dimensional analysis method may provide information about relevant features of the process under investigation by highlighting the most important parameters and, at least, partially showing mutual relationships without the need of solving coupled differential equations. In this specific case, the full theory, and more complex equations, provide a much more precise description. However, the following aspects must be considered: firstly, this level of precision is not always necessary; secondly, precision usually leads to a considerable lengthening of calculations and the need to deal with serious numerical methods; thirdly, dimensional analysis may be applied when precise physical laws for the process are unknown.

Now consider the volume flow rate of blood in an artery (or oil in a tube in a refinery), $\frac{dV}{dt}$, as a function of the pressure drop per unit length ΔP, the radius r, the density ρ, and the viscosity μ. The list of variables for this problem is given in table 5.4.

This problem may also be solved using the Buckingham Π-theorem, but first let us look at it using the Rayleigh algorithm. A generic dimensionless variable will have the form (this is our hypothesis):

$$\Pi = \left(\frac{dV}{dt}\right)^{\mathrm{a}} (\Delta p)^{\mathrm{b}} (r)^{\mathrm{c}} (\rho)^{\mathrm{d}} (\mu)^{\mathrm{e}} \tag{5.24}$$

Table 5.4. Variables for volumetric flow rate.

Variable	Dimension	Description
$\frac{dV}{dt}$	$L^3\,T^{-1}$	Volumetric flow rate (volume/time)
Δp	$ML^{-1}\,T^{-2}$	Pressure drop
r	L	Radius of vessel
ρ	ML^{-3}	Blood density
μ	$ML^{-1}\,T^{-1}$	Blood viscosity

and in terms of dimensions,

$$\begin{aligned} \mathsf{M}^0\mathsf{L}^0\mathsf{T}^0 &= (\mathsf{L}^3\mathsf{T}^{-1})^a(\mathsf{ML}^{-1}\mathsf{T}^{-2})^b(\mathsf{L})^c(\mathsf{ML}^{-3})^d(\mathsf{ML}^{-1}\mathsf{T}^{-1})^e \\ &= \mathsf{M}^{b+d+e}\mathsf{L}^{3a-b+c-3d-e}\mathsf{T}^{-a-2b-e}. \end{aligned} \tag{5.25}$$

We have three equations and five unknowns, so we must choose two free variables. As we are seeking $\frac{dV}{dt}$, we choose a to be one of the free variables. Arbitrarily, we choose e as the second. There are several ways of writing the solutions. But notice that, given the choice we made for free variables, we can write for the base dimensions:

$$\mathsf{M}: b + d = -e$$

$$\mathsf{L}: -b + c - 3d = -3a + e$$

$$\mathsf{T}: -2b = a + e$$

Solving, we obtain: $b = -\tfrac{1}{2}a - \tfrac{1}{2}e$, $c = -2a - e$ and $d = \tfrac{1}{2}a - \tfrac{1}{2}e$. Substituting $a = 2, e = 0$ and $a = 0, e = 2$, we obtain: $a = 2, b = -1, c = -4, d = 1, e = 0$ and $a = 0, b = -1, c = -2, d = -1, e = 2$. Therefore, the two dimensionless groups needed to define the volumetric flow rate (there are n = five variables, described by three base dimensions) are:

$$\Pi_1 = \left(\frac{dV}{dt}\right)^2 \rho/\Delta p\ r^4, \quad \text{and } \Pi_2 = \mu^2/\Delta p\ r^2\rho. \tag{5.26}$$

In this case, the same result is obtained via the Buckingham Π-theorem by solving the two dimensional equations:

$$\begin{aligned} \Pi_1 &= \mathsf{L}^3\mathsf{T}^{-1}(\mathsf{ML}^{-1}\mathsf{T}^{-2})^a(\mathsf{L})^b(\mathsf{ML}^{-3})^c, \text{ and} \\ \Pi_2 &= \mathsf{ML}^{-1}\mathsf{T}^{-1}(\mathsf{ML}^{-1}\mathsf{T}^{-2})^a(\mathsf{L})^b(\mathsf{ML}^{-3})^c; \end{aligned} \tag{5.27}$$

that is, by solving for $\frac{dV}{dt}$, and μ separately, using p, r and ρ as repeat variables.

Either way, the Buckingham Π-theorem tells us there exists a function, f, such that:

$$f\left(\left(\frac{dV}{dt}\right)^2 \rho/\Delta p\ r^4,\ \mu^2/\Delta p\ r^2\rho\right) = 0.$$

By the implicit function theorem, we assume that we can solve this so that there exists a function h:

$$\left(\frac{dV}{dt}\right)^2 \rho/\Delta p\ r^4 = h(\mu^2/\Delta p\ r^2\rho), \tag{5.28}$$

so that: $\frac{dV}{dt} = \frac{r\mu}{\rho}$. (We have omitted the dimensionless constant, which contains the numerical factors.)

Finally, we will look at one of the classic applications of the Buckingham Π-theorem in fluid mechanics; that is, the pressure drop in a long, horizontal, cylindrical tube, or pipe (a conduit). This is similar to the example we have examined, but with many more variables. Engineers and physicists had long known, from experimental observations, that the pressure drop in a pipe—which is the reason a fluid flows through the pipe—is a characteristic of the properties of the pipe. In particular, the pressure drop was seen to be a function of the characteristic length-scale of the pipe, and the interior roughness of the wall of the pipe. The formulation of this problem presented here is taken from [12].

As usual, we set out what we believe to be the key variables for this problem—see table 5.5.

Here, $n = 7$, and the only base dimensions involved here are the usual three mechanical dimensions (M, L and T). Consequently, there are four nondimensional Π-groups to be found. The fundamental hypothesis of the model is that the pressure drop is given by the following function,

$$\Delta p = f(v, d, s, \varepsilon, \rho, \mu). \tag{5.29}$$

The problem is complex, and we will use the Buckingham Π-theorem to analyse this problem, and so we need to decide upon the three repeating variables to be used in deriving the four nondimensional Π-groups. We will choose $\rho(\mathsf{ML}^{-3})$, $d(\mathsf{L})$ and v (LT^{-1}); as they contain the three base dimensions, and do not have base dimensions that are powers or multiples of another repeating parameter (see chapter 3). Importantly, these three repeating variables should not form a non-dimensional group among themselves. If the wrong repeating variables are chosen, the Π-theorem will not give the correct answer. We set up the dimensional equations to form the four dimensionless Π-groups as follows.

1. We first look at the dependent variable, Δp.

$$\begin{aligned}\Pi_1 &= (\rho)^a(v)^b(d)^c\Delta p = (\mathsf{ML}^{-3})^a(\mathsf{LT}^{-1})^b(\mathsf{L})^c(\mathsf{ML}^{-1}\mathsf{T}^{-2}) \\ &= \mathsf{M}^{a+1}\mathsf{L}^{-3a+b+c-1}\mathsf{T}^{-b-2}.\end{aligned} \tag{5.30}$$

The solutions are: $a = -1$, $b = -2$ and $c = 0$, which yield, $\Pi_1 = \Delta p/\rho v^2$.

2. The Π-group involving the viscosity,

Table 5.5. Variables for generalized flow in a pipe.

Variable	Dimension	Description
Δp	$\mathsf{ML}^{-1}\,\mathsf{T}^{-2}$	Pressure drop in pipe (the dependent variable)
v	LT^{-1}	Area average velocity through pipe
d	L	Inner diameter of pipe
s	L	Length of pipe
ε	L	Roughness element length (e.g. standard deviation from smoothness)
ρ	ML^{-3}	Density of fluid
μ	$\mathsf{ML}^{-1}\,\mathsf{T}^{-1}$	Dynamic viscosity of fluid

$$\Pi_2 = (\rho)^d(v)^e(d)^f\mu = (\mathrm{ML^{-3}})^d(\mathrm{LT^{-1}})^e(\mathrm{L})^f(\mathrm{ML^{-1}T^{-1}})$$
$$= \mathrm{M}^{d+1}\mathrm{L}^{-3d+e+f-1}\mathrm{T}^{-e-1}. \tag{5.31}$$

The solutions are: $d = -1$, $e = -1$ and $f = -1$, which yield, $\Pi_2 = \mu/d\rho v$. Then for the two Π-groups involving a length:

$$\Pi_3 = (\rho)^g(v)^h(d)^i s = (\mathrm{ML^{-3}})^g(\mathrm{LT^{-1}})^h(\mathrm{L})^i(\mathrm{L})$$
$$= \mathrm{M}^g\mathrm{L}^{-3g+h+i+1}\mathrm{T}^{-h}. \tag{5.32}$$

The solutions are: $g = 0$, $h = 0$ and $i = -1$, which yield, $\Pi_3 = s/d$.

$$\Pi_4 = (\rho)^j(v)^k(d)^l\varepsilon = (\mathrm{ML^{-3}})^j(\mathrm{LT^{-1}})^k(\mathrm{L})^l(\mathrm{L})$$
$$= \mathrm{M}^j\mathrm{L}^{-3j+k+l+1}\mathrm{T}^{-k}. \tag{5.33}$$

The solutions are: $j = 0$, $k = 0$ and $l = -1$, which yield, $\Pi_4 = \varepsilon/d$.

Therefore, our function for the four nondimensional parameters in this problem may be written as:

$$\Delta p/\rho v^2 = F_1\left(\mu/d\rho v, \frac{s}{d}, \frac{\varepsilon}{d}\right). \tag{5.34}$$

This is as far as the Buckingham Π-theorem, or dimensional analysis can take us in this problem. From here, we have to use experimental data and physical insight to further specify equation (5.34). For example, we can immediately simplify equation (5.34) by noting that the pressure drop is expected to vary in direct proportion to the length of the pipe, s (see the analysis for the pressure drop in a blood vessel above). We should then take care in our dimensional analysis to isolate the length, so that it occurs in only one nondimensional parameter, here s/d (see chapter 3). We can then rewrite equation (5.34) isolating the known dependence:

$$\Delta p/\rho v^2 = \frac{s}{d}F_2\left(\frac{\mu}{\rho v d}, \frac{\varepsilon}{d}\right). \tag{5.35}$$

This is a reasonable representation for this function, and accurately mimics experimental data [13]. However, further simplification is possible if we adopt more conventional notation; in particular, the nondimensional Reynolds number is defined as $\frac{\rho v d}{\mu}$. And inverting a dimensionless group is permitted, giving:

$$\Delta p/½\rho v^2 = \frac{s}{d}F_3\left(\frac{\rho v d}{\mu}, \frac{\varepsilon}{d}\right). \tag{5.36}$$

As per convention in fluid mechanics, we now define a factor f equal to the arbitrary function we have derived from our dimensional analysis as follows:

$$f = F_4\left(\frac{\rho v d}{\mu}, \frac{\varepsilon}{d}\right) = F_4\left(\mathrm{Re}, \frac{\varepsilon}{d}\right). \tag{5.37}$$

The function F_4 is formally termed the friction factor for pipe flow. This is also termed the Darcy–Weisbach friction factor, and is widely used in engineering. People have been making detailed measurement of the flow of fluids through pipes for two centuries. The shape of the three-variable function given in equation (5.37) (describing the relation between the nondimensional groups of variables we have generated from our dimensional analysis) has been carefully mapped out. Such data is shown in [13], and it is seen that equation (5.37) is capable of rationalizing observed experimental data. Interestingly, the experimental data also demonstrates that roughness of the interior surface of the pipe has a significant influence on the dynamical properties of the flow (see section 7.1). Consequently, it is not possible to ignore the term $\frac{\varepsilon}{d}$ in equation (5.34) as being a negligible contribution as might be expected. An example of such experimental data may be found in the well-known Moody diagram (see figure 5.4).

Similitude requires that if mechanically similar flow is to take place in two pipes, they must have a geometrically similar form, and have similar interior wall surfaces. The first requirement is met with pipes of circular cross-section. The second requirement is satisfied by maintaining a constant ratio of the pipe radius to the magnitude of any surface irregularities (what we have defined as $\frac{\varepsilon}{d}$). It is essential, therefore, that the materials producing the roughness should be similar. Thus, we bring dimensional analysis into the world of industrial metrology.

Curves, such as those displayed in figure 5.4, are all determined from experimental data. From such data, scaling rules or laws may be deduced, which can be used in a less rigorous environment. They provide precision to a few percent.

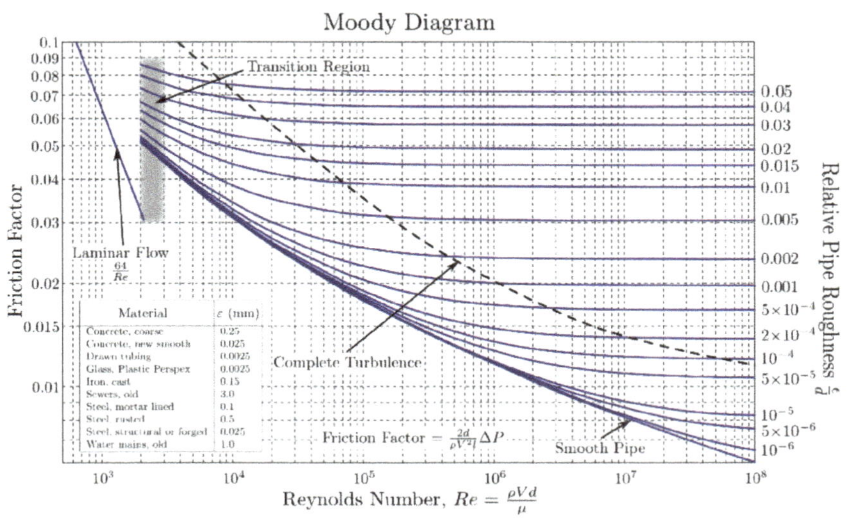

Figure 5.4. Moody diagram showing the Darcy–Weisbach friction factor (equation (5.37)) plotted against the Reynolds number for various relative roughness ε/d. This 'Moody diagram' image has been obtained by the author from the Wikimedia website where it was made available under a CC BY-SA 4.0 license. It is included within this article on that basis. It is attributed to S Beck and R Collins, University of Sheffield.

Fox *et al* [12] have fitted data to produce the following scaling law, in the region of the cross-over from laminar to turbulent flow,

$$\frac{1}{\sqrt{f}} \approx -1.8 \log_{10}\left\{\frac{6.9}{\mathrm{Re}} + \left(\frac{\varepsilon}{d}\frac{1}{3.7}\right)^{1.11}\right\}.$$

Without attempting to solve the Navier–Stokes equations, we have derived a useful relation among a number of easily measured variables. That is, by using experimental data in the context of dimensional analysis, we can determine the approximate underlying function. The fractional power of 1.11 reminds us that this is a scaling rule, and not an equation derived from first principles.

References

[1] Rayleigh L 1915 The principle of similitude *Nature* **95** 66–8
[2] Bridgman P W 1931 *Dimensional Analysis* 2nd edn (New Haven, CT: Yale University Press)
[3] Sedov L I 1959 *Similarity and Dimensional Analysis in Mechanics* (New York: Academic)
[4] Taylor E S 1974 *Dimensional Analysis for Engineers* (Oxford: Clarendon)
[5] Barenblatt G I 1996 *Scaling, Self-Similarity, and Intermediate Asymptotics* (Cambridge: Cambridge University Press)
[6] McMahon T A and Bonner J T 1983 *On Size and Life* (New York: Scientific American Library)
[7] Tennekes H 1997 *The Simple Science of Flight* (Cambridge, MA: MIT Press)
[8] de Jong F J 1967 *Dimensional Analysis for Economists* (Amsterdam: North Holland)
[9] BIPM, IEC, IFCC, ILAC, ISO, IUPAC, IUPAP, and OIML Int. Vocabulary of Metrology—basic and general concepts and associated terms (VIM) 2008 https://bipm.org/utils/common/documents/jcgm/JCGM_200_2012.pdf
[10] Buckingham E 1914 On physically similar systems; illustrations of the use of dimensional equations *Phys. Rev.* **4** 345–76
Buckingham E 1915 The principle of similitude *Nature* **96** 396–7
Buckingham E 1915 Model experiments and the forms of empirical equations *Trans. of the American Society of Mechanical Engineers* **37** 263–296
[11] Lira I 2013 Dimensional analysis made simple *Eur. J. Phys.* **34** 1391–401
[12] Fox R W, McDonald A T and Pritchard P J 2015 *Introduction to Fluid Mechanics* 9th edn (Hoboken, NJ: Wiley)
[13] White F M 2016 *Fluid Mechanics* 9th edn (New York: McGraw-Hill)

IOP Publishing

Dimensional Analysis
The great principle of similitude
Jeffrey H Williams

Chapter 6

Scaling and similitude

'Everything—yes, everything—is a harmonic oscillator'
(Title of an article in *Wired*, 18th July 2016)

Scaling is a notion that should really be second nature to those seeking to solve problems; for example, one should always be conscious of how a dependent quantity, y, scales with variations of an independent quantity, x. Similitude is related to scaling: suppose we scaled a parameter, x, by a factor, C. If we now wished to maintain the same value of quantity, y, how would we need to rescale the other independent parameters? Put another way: if we measure an invariant quantity, such as the speed of light, we could do no better than measure 299 792 458 metres per second. Then, if we wished to change our length and time units so that the speed of light in the new system of units is 1, we have to rescale a great many other quantities. However, the invariance of physics, and the requirement of a consistent system of units is something we usually do not think about consciously when seeking to solve problems—it is ingrained and becomes second nature.

This chapter presents an outline of how one can use dimensional analysis to arrive at a physical scaling relation through inspection. Such combinations are not always easy to see, and you may wonder if you have spotted them all. Fortunately, the Buckingham Π-theorem can also be used to analyse those cases where the Rayleigh algorithm has limitations.

6.1 Astronomy and the music of the spheres

Johannes Kepler (1571–1630) was a German astronomer, mathematician and astrologer. He is a key figure in the 17th Century scientific revolution, best known for his laws of planetary motion, and his books: *Astronomia nova*, *Harmonices Mundi* and *Epitome Astronomiae Copernicanae*. These works laid the foundations

doi:10.1088/978-0-7503-3655-0ch6

for Newton's theory of universal gravitation. To finance his studies, Kepler was a mathematics teacher at a seminary school in Graz, Austria. Later he became an assistant to the astronomer Tycho Brahe in Prague, and eventually Imperial Mathematician to the Holy Roman Emperor Rudolf II, and his two imperial successors Matthias and Ferdinand II. Additionally, he did fundamental work in the field of optics; invented an improved version of the refracting telescope, and is mentioned in the telescopic discoveries of his contemporary Galileo Galilei.

Johannes Kepler lived in an era when there was no clear distinction between astronomy and astrology, but there was a strong division between astronomy (a branch of mathematics within the liberal arts) and physics (a branch of natural philosophy). Kepler described his new astronomy as '*celestial physics*', as '*an excursion into Aristotle's Metaphysics*', and as '*a supplement to Aristotle's On the Heavens*'; he began the transformation of the ancient tradition of physical cosmology by treating astronomy as part of a universal mathematical physics. Kepler can very well be considered as an exemplar of the dawn of modern science. Although there is still more than a hint of metaphysics in Kepler's physics.

Given the central place accorded to astronomy in 17th Century Europe, we will begin our exploration of the application of dimensional analysis by looking at some of the earliest major physical discoveries made using mathematics. In astronomy, Kepler's laws of planetary motion are three laws or rules describing the motion of planets around the Sun, published between 1609 and 1619. They are often held up as the first true 'law of physics'. These three laws improved upon the heliocentric theory of Nicolaus Copernicus; replacing its circular orbits and epicycles with elliptical trajectories, and explaining how planetary velocities vary. The three laws state:

- The orbit of a planet is an ellipse with the Sun at one of the two *foci*.
- A line segment joining a planet and the Sun sweeps out equal areas during equal intervals of time.
- The ratio of the square of an object's orbital period with the cube of the semi-major of its orbit is the same for all objects orbiting the same primary.

Johannes Kepler's third law—sometimes referred to as the law of harmonies—compares the orbital period and radius of orbit of a planet to those of other planets. Unlike Kepler's first and second laws that describe the motion characteristics of a single planet, the third law makes a comparison between the dynamics of different planets. The comparison being made is that the ratio of the squares of the periods (time, T) to the cubes of their average distances from the Sun (distance, R) is the same for every one of the planets. Kepler enunciated this law in 1619; it is an attempt to determine what he viewed as the 'music of the spheres'. But Kepler's third law actually provides an accurate description of the period and distance for a planet's orbit about the Sun. Additionally, the law that describes the R^3/T^2 ratio for the planets' orbits also describes the R^3/T^2 ratio for any satellite (whether a moon or a man-made satellite) orbiting about any planet.

On discovering this scaling rule operating through the Solar System, and being able to largely explain the variation in the behaviour between any pair of planets using Pythagoras' concept of musical intervals, Kepler commented '*I first believed*

I was dreaming... But it is absolutely certain and exact that the ratio which exists between the period times of any two planets is precisely the ratio of the 3/2 *power of the mean distance*' (from *Harmonies of the World* by Kepler, 1619). Figure 6.1 demonstrates the scaling rule discovered by Kepler (it is based on modern data, and contains those planets discovered since Kepler's time). The scaling law discovered by Kepler collapses a set of apparently disparate data onto a single scaled rule or law.

Johannes Kepler proposed in his law of harmonies, that the ratio of the mean radius of the orbit cubed (R^3) to the period of orbit squared (T^2) is a constant—with the same value, k—for all the planets that orbit the Sun. The observational data of Kepler's day for the planets suggested the following average ratio: $R^3/T^2 \approx 7.64$–7.43×10^{-6} AU cubed per day squared; the AU is the Astronomical Unit, the average distance from the Earth to the Sun. By equating the same units of force, Isaac Newton was able to combine the law of universal gravitation with the circular motion principles to show that if the force of gravity provides the centripetal force for the planets' nearly circular orbits, he could predict values of R^3/T^2 that agreed with Kepler's measurements.

Consider a planet with mass, M_{planet}, in a nearly circular orbit about a sun of mass, M_{Sun}, moving at a velocity, v, and at a distance, R. The net centripetal force (F_{net}) acting upon the orbiting planet is given by

$$F_{\text{net}} = M_{\text{planet}}\, v^2/R; \tag{6.1}$$

dimensionally, this is: $[F_{\text{net}}] = \mathrm{MLT}^{-2} = \mathrm{M(LT^{-1})^2L^{-1}} = \mathrm{MLT}^{-2}$. This net centripetal force is the result of the gravitational force (F_{grav}) that attracts the planet towards the Sun, and can be represented by Newton's law of universal gravitation, with gravitational constant, G,

$$F_{\text{grav}} = GM_{\text{planet}}M_{\text{Sun}}/R^2. \tag{6.2}$$

Equating these two forces (they have the same dimensionality), F_{grav} and F_{net}, we have:

$$M_{\text{planet}}v^2/R = GM_{\text{planet}}M_{\text{planet}}M_{\text{Sun}}/R^2. \tag{6.3}$$

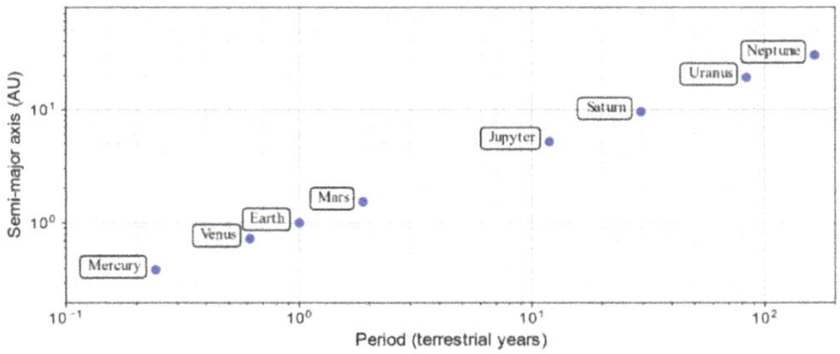

Figure 6.1. Log–log plot of the semi-major axis (in Astronomical Units) versus the orbital period (in terrestrial years) for the eight planets of the Solar System. Reproduced from https://commons.wikimedia.org/wiki/File:Solar_system_planets_a(AU)_vs_period(terrestrial_years).svg. Image stated to be in the public domain. It is included within this article on that basis.

Since the velocity of an object in nearly circular orbit can be approximated as $v = (2\pi R)/T$, or $v^2 = 4\pi^2 R^2/T^2$, substitution of the expression for v^2 into the equation above yields

$$(M_{\text{planet}}4\pi^2R^2)/RT^2 = GM_{\text{planet}}M_{\text{Sun}}/R^2. \tag{6.4}$$

The equation can be transformed to

$$R^3/T^2 = GM_{\text{planet}}M_{\text{Sun}}/M_{\text{planet}}4\pi^2. \tag{6.5}$$

The mass of the planet can then be eliminated from the numerator and the denominator of the RHS yielding:

$$R^3/T^2 = GM_{\text{Sun}}/4\pi^2 \approx 7.496 \times 10^{-6}\text{AU}^3/(\text{days})^2. \tag{6.6}$$

The RHS of this equation will be a constant for every planet regardless of the planet's mass. This constant can be determined from the slope of the line in figure 6.1. And the ratio R^3/T^2 would be the same value for all planets, provided the force that holds the planets in their orbits is the force of gravity. Newton's law of gravitation predicts results that were consistent with known planetary data, and provided a theoretical explanation for Kepler's Law. This is a *tour de force* in the application of the principle of similitude. And if it happens in our Solar System, it happens everywhere.

Isaac Newton's proof, and explanation of Kepler's third law could only have come about because Newton was thinking in terms of the dimensions of the physical properties and phenomena involved in this particular problem. Newton and Kepler both knew that they were dealing with forces; that is, something unseen that causes action at a distance. Dimensionally, all forces are MLT^{-2}; or you could say that they are energy (E) directed in a particular direction; that is, $(\partial E/\partial L) = \mathsf{ML}^2\mathsf{T}^{-2}/\mathsf{L}$. Newton was therefore able to equate two different expressions for forces, and derive a new insight into the way Nature functions.

6.1.1 Dimensional analysis of Kepler's third law

How does the period (P) of an orbiting mass depend on the radius of its orbit? For simplicity, we have a small planetary mass, m, orbiting a larger mass, M_0, at a distance, r. Here, the variables are: the period of the orbit, P, the central mass, M_0, the radius of the orbit, r, and the Gravitational constant, G (not 'little' g). Table 6.1 contains a description of these variables.

We first assume some form for the property of interest, the period of the orbit; for example, $P = k\,(M_0)^a\,(r)^b\,(G)^c$ (again, we have a dimensionless constant of proportionality, k). Inserting the dimensions of the variables gives,

$$\mathsf{T} = \mathsf{M}^a\mathsf{L}^b(\mathsf{M}^{-1}\mathsf{L}^3\mathsf{T}^{-2})^c = \mathsf{M}^{a-c}\mathsf{L}^{b+3c}\mathsf{T}^{-2c}. \tag{6.7}$$

The solutions are: $a = -½$, $c = -½$ and $b = 3/2$. This allows us to write

$$P = kM_0^{-1/2}r^{3/2}G^{-1/2}, \text{ or } P^2 = (k^2/GM_0)r^3 \rightarrow r^3/P^2 = GM_0/k^2. \tag{6.8}$$

The constant may be found by equating the two forces; $k^2 = 4\pi^2$.

Table 6.1. Variables for planetary motion.

Variable	Dimension	Description
P	T	Period of orbit
m	M	Mass of planet
M_0	M	Mass of star
r	L	Distance between the two masses
G	$M^{-1}L^3T^{-2}$= {$MLT^{-2}.L^2.M^{-2}$}	Newton's gravitational constant (6.674 30(15) × 10^{-11} m^3 kg^{-1} s^{-2})

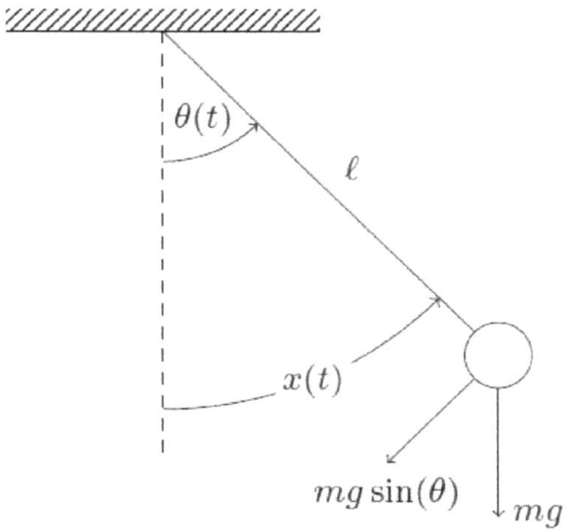

Figure 6.2. A schematic of the essential details of a simple pendulum.

6.2 Dimensional analysis of the pendulum: the first precision measuring device

At its simplest, a pendulum is merely a heavy weight or 'bob' suspended at the end of a long thin wire from a fixed pivot, so that it may swing freely (see Figure 6.2). When the bob is displaced from its static, vertical, or resting equilibrium position so that it is higher or further from the centre of the Earth, gravity will exert a pulling force to return the bob to the initial resting equilibrium, or vertical position. But, by the time that the bob reaches its starting position the bob is moving, and momentum causes the bob to move through the initial, vertical position and it swings upwards until gravity exerts a restoring force to bring it back. These competing forces of momentum and gravity cause the bob to oscillate about the original equilibrium position; that is, the pendulum is observed to swing back and forth.

The time for one complete cycle of the bob, a left swing plus a right swing, is called the period of the pendulum. If you observe the swinging motion or oscillations of the pendulum, you will see that the arc described by the motion of the bob is constant over the time-scale of most observations. This apparent, constant period of oscillation was used soon after its discovery as a means of measuring the force of gravity that Isaac Newton had announced was all pervading, and which held the Universe together. However, as with most things in physics, there are qualifications to describing a pendulum as a simple harmonic oscillation. For a pendulum to behave as a perfect gravity pendulum, the pivot from which the weight is suspended must be frictionless, the wire suspending the bob must not have any mass, and not be able to stretch or deform. In addition, the bob should not suffer any drag or fictional force as it moves through the air. These conditions of perfection are impossible to achieve under everyday conditions, and eventually all pendulums will run down naturally, and the bob will eventually return to its static, vertical position.

The variables involved in the dynamics of the pendulum are given in table 6.2. But before applying dimensional analysis, let us look at the problem in classical mechanics; that is, we formulate and solve the ordinary differential equations that model the dynamics. By Newton's second law of motion,

$$\mathrm{m}\frac{d^2x(t)}{dt^2} = -\mathrm{mg}\sin(\theta(\mathrm{t})). \tag{6.9}$$

However, $x(t) = l\,\theta(t)$, thus,

$$\frac{d^2\theta}{dt^2} = -\frac{g}{l}\sin(\theta). \tag{6.10}$$

We will consider small amplitudes of oscillation, so θ remains small. Then, $\sin\theta = \theta + 0(\theta^3)$ as $\theta \to 0$. Neglecting terms of order θ^3 gives

$$\frac{d^2\theta}{dt^2} = -\frac{g}{l}\theta. \tag{6.11}$$

The general solution to this differential equation is

$$\theta(t) = A\cos(\omega t + \alpha), \tag{6.12}$$

where $\omega = \sqrt{g/l}$, and A and α are constants. Now,

$$\theta\left(t + \frac{2\pi}{\omega}\right) = \cos\left\{\omega\left(t + \frac{2\pi}{\omega}\right) + \alpha\right\} = \cos(\omega t + \alpha + 2\pi) = \theta(t). \tag{6.13}$$

Table 6.2. Variables for a pendulum.

Variable	Dimension	Description
T_0	T	Period of oscillation
m	M	Mass of bob
l	L	Length of pendulum
g	LT^{-2}	'Little' g this time. Acceleration due to gravity (9.806 65 m s^{-2})
θ	-	Angle of deflection from the vertical in radian

Thus,

$$T_0 = \frac{2\pi}{\omega} = 2\pi\sqrt{l/g}, \tag{6.14}$$

where T_0 is the period of oscillation, neglecting terms to $0(\theta^3)$.

Having analysed the problem in some detail, let us now consider a dimensional analysis. We will again calculate the period of oscillation. The variables are given in table 6.2. Assume that there exists a function, f, such that $T_0 = f(l, g, m, \theta)$; this comes from assuming: $T_0 \propto (l)\,(g)\,(m)\,(\theta)$. The three base dimensions involved in this problem are also given in table 6.2.

We consider a product of powers of l, g, m, θ such that the product has the same dimension as T_0: $[T_0] = [l^a g^b m^c \theta^d] = [l]^a[g]^b[m]^c[\theta]^d$, which gives us the dimensional equation (note: we have written $[\theta] = 1$, because angle is dimensionless):

$$\mathsf{T} = \mathsf{L}^a(\mathsf{LT}^{-2})^b\mathsf{M}^c\,1^d. \tag{6.15}$$

Equating the exponents of M, L and T, and we find (dimensionless θ disappears):

$$\mathsf{M}:\ c = 0$$

$$\mathsf{L}:\ a + b = 0$$

$$\mathsf{T}:\ -2b = 1.$$

Thus, $a = ½$, $b = -½$ and $c = 0$, and we observe that the period of oscillation is independent of the mass of the bob. Thus, T_0 and $\sqrt{(l/g)}$ have the same dimension (T). Now, consider $T_0 = f(l, g, m, \theta)$, and divide both sides by $\sqrt{(l/g)}$:

$$T_0/\sqrt{(l/g)} = f(l, g, m, \theta)/\sqrt{(l/g)}. \tag{6.16}$$

Both sides of Eq. (6.16) are now dimensionless. Let

$$\sqrt{(g/l)}f(l, g, m, \theta) = F(l, g, m, \theta). \tag{6.17}$$

Thus,

$$\sqrt{(g/l)}T_0 = F(l, g, m, \theta). \tag{6.18}$$

Changing the unit of measurement for m, l or g will change the value of the variable, but it will not change the LHS of the above equation because it is dimensionless. Hence, $F(l, g, m, \theta)$ must be independent of m, l and g. Thus, $\sqrt{(g/l)}\,T_0 = F(\theta)$, and therefore,

$$T_0 = F(\theta)\sqrt{(l/g)}. \tag{6.19}$$

Since there is no damping, $F(\theta)$ is an even function of θ. Expand $F(\theta)$ in a Taylor series:

$$F(\theta) = F(0) + \theta F'(0) + 1/2\theta^2F''(0) + \mathrm{O}(\theta^3). \tag{6.20}$$

Since $F(\theta)$ is an even function of θ, $F'(0) = 0$, $F'''(0) = 0$, …, and therefore,

$$F(\theta) = F(0) + 1/2\theta^2 F''(0) + O(\theta^4). \tag{6.21}$$

Thus, to first order in θ,

$$T_0 = F(0)\sqrt{(l/g)}. \tag{6.22}$$

If we compare the results of our two calculations of the period of the pendulum; that is, compare equations (6.14) and (6.22), we see that dimensional analysis gives the same answer as calculus except for the numerical factor of 2π. However, the expression $F(0)$ can be obtained experimentally using a pendulum of any length. We see that dimensional analysis gives the correct functional form of the solution in terms of the physical quantities. It does not, however, give constant factors; these factors have to be obtained experimentally.

We will now look at the same problem with the Buckingham Π-theorem. For this, we form dimensionless groups from the variables given in table 6.2. The $n = 5$ variables contain $m = 3$ base dimensions. The Buckingham Π-theorem says that we can form $n - m = 2$ independent dimensionless groups from this set. One original variable, θ_0, is already dimensionless so we start with $\Pi_1 \equiv \theta_0$. Only one original variable, m, contains the dimension of mass. Therefore, no dimensionless group can contain m. (To be dimensionless, the group has to contain another variable that cancels the unit of mass contributed by m; hence the period is independent of m, M.) We form the second independent dimensionless variable using g, l and T_0. The units of g show us how to form the group: Cancel the length in g with l^{-1}, and cancel the time with T_0^2, to make $\Pi_2 \equiv gT_0^2/l$. The two groups are then (Figure 6.3 summarises this process):

$$\Pi_1 = \theta_0, \text{ and } \Pi_2 = gT_0^2/l. \tag{6.23}$$

The most general relation using these nondimensional groups is

$$f(\Pi_1, \Pi_2) = 0, \tag{6.24}$$

where f is a dimensionless function. The two-parameter function f is equivalent to a family of one-parameter functions, f_{Π_1}, indexed by Π_1. The general form equation (6.24) is then

$$f_{\Pi_1}(\Pi_2) = 0. \tag{6.25}$$

Where f was a one-parameter function, we transformed $f(\Pi_1) = 0$ to $\Pi_1 = \Pi$. Here, we transform $f_{\Pi_1}(\Pi_2) = 0$ to

$$\Pi_2 = f(\Pi_1), \tag{6.26}$$

where $f(\Pi_1)$ is the zero of $f_{\Pi 1}$. (This transformation is valid when $f_{\Pi 1}$ has only one physically valid zero, which we assume for simplicity.) We have extracted Π_2 from f. Alternatively, we could have written $f(\Pi_1, \Pi_2) = f_{\Pi_2}(\Pi_1)$, and defined $f(\Pi_2)$ as the zero of $f\,\Pi_2$. Then, we could have extracted Π_1:

$$\Pi_1 = f(\Pi_2). \tag{6.27}$$

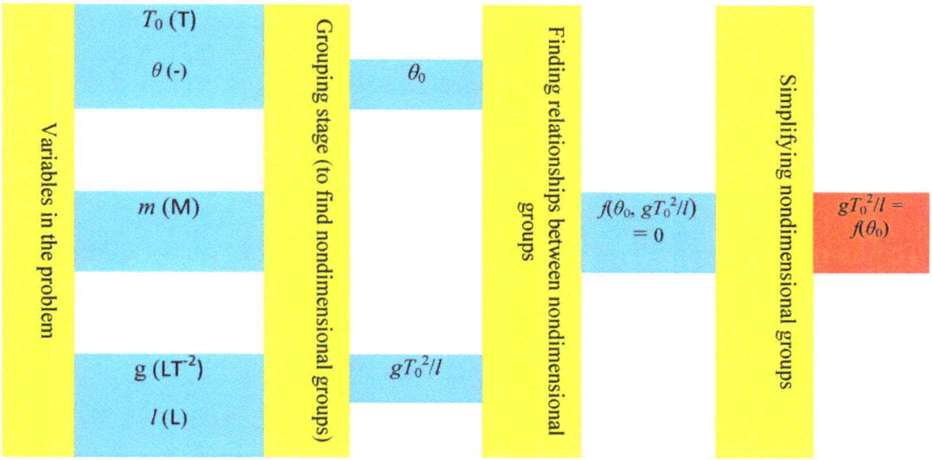

Figure 6.3. Schematic summary of the dimensional analysis for the period of a pendulum. We choose five relevant variables (see table 6.2). They go into the grouping stage, which manipulates the inputs (five) and the number of independent dimensions in the inputs (three), and generates two dimensionless Π-groups. The relation finder produces the most general relation using the two groups. The simplifier transforms the relation into a form more convenient for solving for T_0.

We choose which form to use by the location of the desired variable. If it belongs to Π_2, we use equation (6.26). If it belongs to Π_1, we use equation (6.27). If it belongs to both, then we have to think about the variables used to derive the dimensionless groups. If it belongs to neither, then we have too few starting variables and dimensionless groups, and redo the analysis. If we wish to solve for the period, which is contained in $\Pi_2 = gT_0^2/l$, so we use equation (6.26). In terms of the original variables, equation (6.26) is $gT_0^2/l = f(\theta)$, or

$$T_0 = f(\theta)\sqrt{\frac{1}{g}}. \tag{6.28}$$

Again, equation (6.28) is as seen in equation (6.14) and equation (6.22). We do not know the function $f(\theta)$; we can determine it by experiment: We release a pendulum at various values of θ, and measure $T_0(\theta)$. Then, f is

$$f(\theta) = T_0(\theta)\sqrt{\frac{g}{l}}. \tag{6.29}$$

We do not have to repeat the experiments for different l (for another pendulum) or g (on top of a mountain), because f is general. In the small-amplitude limit ($\theta \to 0$), we can simplify equation (6.16):

$$T_0 = f(0)\sqrt{\frac{l}{g}}, \tag{6.30}$$

where $f(0)$ is a constant. In speaking of $f(0)$, we are assuming that $\lim_{x\to 0} f(x)$ exists. Dimensional analysis cannot provide this insight—we have to consider the physics.

We estimate the oscillation time by estimating the acceleration and the oscillation distance. The pendulum feels a force, $F \sim mg\sin\theta$ (see figure 6.2), which makes it accelerate at $a \sim g\sin\theta$. For small θ, the acceleration is $a \sim g\theta$. In time t, it travels a distance, $at^2 \sim g\theta t^2$. It needs to travel a distance $d \sim l\theta$ (neglecting constants) to complete a cycle, so

$$g\theta t^2 \sim l\theta. \tag{6.31}$$

The amplitude θ cancels, which is the physical equivalent of the mathematical statement that $f(0)$ exists. We even derive an estimate for t,

$$t \sim \sqrt{\frac{l}{g}}. \tag{6.32}$$

Alternatively, we can estimate a typical velocity for the bob; and from the velocity, estimate the period. The maximum potential energy is $PE \sim mgh$, where the change in height is $h = l(1 - \cos\theta) \sim l\theta^2$. The maximum kinetic energy is equal to the maximum potential energy. So, the maximum velocity is given by $mv^2 \sim mgl\theta^2$. (This relation equates two energies, potential and kinetic; we are invoking dimensional homogeneity.) The maximum velocity is then $v \sim \theta\sqrt{gl}$; we use this value as an estimate for a typical velocity. The time to complete a cycle is $t \sim d/v \sim \sqrt{l/g}$, as we found using the acceleration method.

Let us now add a perturbation to our pendulum: We will consider how air resistance perturbs the motion of the bob. This is equivalent to coupling a drag force, R, to the dynamics of the bob—that is, the period of oscillation will be perturbed. We add a force, R, $[R] = \mathsf{MLT}^{-2}$, to the list of variables in table 6.2. And our model equation relating these variables to the period of the pendulum now becomes

$$[T_0] = [l^a m^b g^c \theta^d R^e] = [l]^a[m]^b[g]^c[\theta]^d[R]^e. \tag{6.33}$$

An ordering of the fundamental dimensions involved, yields the dimensional equation:

$$\mathsf{T} = \mathsf{L}^a\mathsf{M}^b(\mathsf{LT}^{-2})^c(\mathsf{MLT}^{-2})^e \rightarrow \mathsf{T} = \mathsf{M}^{b+e}\mathsf{L}^{a+c+e}\mathsf{T}^{-2c-2e}. \tag{6.34}$$

From which we obtain the following system:

$$\mathsf{M}:\ b + e = 0$$

$$\mathsf{L}:\ a + c + e = 0$$

$$\mathsf{T}:\ -2c - 2e = 1.$$

As there are four variables and only three equations; we cannot determine the solution uniquely, but can solve for three variables in terms of the fourth (free variable). Choose any of the variables to be the free variable; for example, with b as free variable, we obtain $a = ½$; $b = b$; $c = -½+b$; $e = -b$. So,

$$T_0 = kl^{1/2}m^b g^{-1/2+b}\,\theta^d R^{-b}. \tag{6.35}$$

Now we can group the variables into groups of dimensionless products, and obtain the following model:

$$T_0 = k\left(\frac{l}{g}\right)^{1/2}\left(\frac{mg}{R}\right)^{b}\theta^{d}. \tag{6.36}$$

Note that solving the system of equations for c instead of b would give us $a = 1/2$; $b = 1/2 + c$; $c = c$; $e = -½-c$. This gives another, equally valid model:

$$T_0 = kl^{1/2}m^{1/2+c}g^{c}\theta^{d}R^{-1/2-c} = k\left(\frac{lm}{R}\right)^{1/2}\left(\frac{mg}{R}\right)^{c}\theta d. \tag{6.37}$$

The example demonstrates that the selected model may not be completely determined. In such cases; that is, when the use of the Rayleigh algorithm requires the use of free variables, the Buckingham Π-theorem becomes essential.

As seen previously, in the Π-technique we group the variables into dimensionless products $\Pi_1, \Pi_2, \ldots \Pi_{n-m}$ and obtain the solution in the form $f(\Pi_1, \Pi_2, \ldots. \Pi_{n-m}) = 0$ or, equivalently, if solving for one of the products (say the first one) $\Pi_1 = g(\Pi_2, \ldots \Pi_{n-m})$. In case there is just one product Π_1, this last equation becomes Π_1 = constant. The dimensionless products in the case of the simple pendulum are $\Pi_1 = T_0^2 g/l$; and $\Pi_2 = \theta$; and in the case of the pendulum with air resistance are $\Pi_1 = T_0^2 g/l$; $\Pi_2 = \theta$; and $\Pi_3 = mg/R$.

To apply the Buckingham Π-theorem to the problem of a pendulum with air resistance, start with $l^a m^b g^c \theta^d R^e T_0^f$. Considering the dimensions, you obtain: $L^a M^b (LT^{-2})^c (MLT^{-2})^e T^f \rightarrow M^{b+e} L^{a+c+e} T^{-2c-2e+f}$. The three dimensional equations are (there are five unknown quantities in these three equations):

$$M: b + e = 0,$$

$$L: a + c + e = 0,$$

$$T: -2c - 2e + f = 0.$$

Chose f to be one of the required free variables; d is also a free variable in this problem. When choosing e for the remaining free variable, and solving for a; b and c; one obtains: $a = -½f$, $b = -e$; and $c = -e + ½f$. Substituting back into $l^a m^b g^c \theta^d R^e T_0^f$; one obtains:

$$l^{-1/2f} m^{-e} g^{-e+1/2f} \theta^d R^e T_0^f = \theta^d (m^{-1}g^{-1}R)^e (l^{-1/2}g^{1/2}T_0)^f. \tag{6.38}$$

This yields three dimensionless products $\Pi_1 = \theta$; $\Pi_2 = m^{-1}g^{-1}R = \frac{R}{mg}$; and $\Pi_3 = l^{-1/2} g^{1/2} T_0 = \frac{\sqrt{gt}}{\sqrt{l}}$. A feasible model can be obtained in the form $\Pi_3 = k\,\Pi_1^{\,d}\,\Pi_2^{\,e}$; solving for T_0, you obtain:

$$T_0 = k\left(\frac{l}{g}\right)^{1/2}\left(\frac{R}{mg}\right)^{e}\theta^{d}. \tag{6.39}$$

The only way to proceed from here is by experiment.

6.3 Harmonic oscillations

The law governing simple harmonic motion was first stated by the English *savant* Robert Hooke (1635–1703) in 1660 as *Ut tensio, sic vis* (*As the extension, so the force*). However, if the amplitude of the oscillations is large, then the harmonic approximation to the motion cannot be made, and the equations governing the motion of, for example, the pendulum are more difficult to solve, and there is no simple relationship between the length of the pendulum and gravity. For swings of small amplitude, the period or frequency of swing is approximately the same for different size swings; that is, the period is independent of amplitude. This property, termed isochronism, is the reason pendulums are so useful for time-keeping and was first identified by Galileo. Successive swings of the pendulum take the same amount of time. Galileo first employed free swinging pendulums in simple timing applications, such as an early metronome for musicians, and a doctor friend used it as a time-piece to take a patient's pulse. In 1641, Galileo produced a design for a pendulum clock. However, the first true pendulum clock was built in 1656 by the Dutch *savant* and inventor Christiaan Huygens (1629–1695). The precision of this first pendulum clock was a great improvement over existing mechanical clocks; increasing precision from about ±15 min per day to around ±15 seconds per day. With the invention of temperature-compensated pendulums in 1721, errors in precision pendulum clocks fell to a few seconds per week. The world was then ready for the advances in international time-keeping and surveying, which would characterise the 18th Century.

During his expedition to Cayenne, French Guiana (northeast coast of South America), in 1671, the French explorer and astronomer Jean Richer (1630–1696) found that a pendulum clock was slower by two and a half minutes per day at Cayenne than the same pendulum clock at Paris. From this he deduced that the force of gravity was lower in French Guiana. In 1687, Isaac Newton in his *Principia Mathematica* explained that this variation in the force of gravity arose because the Earth was not a true sphere, but slightly oblate (flattened at the poles and bulging at the equator). This imperfection in the Earth's roundness coupled with the effect of centrifugal force due to its rotation, caused gravity to increase with latitude. As a consequence, portable pendulums began to be taken on voyages to distant lands, as precision gravimeters to measure the value of gravity at different points on the Earth's surface. These measurements were subsequently used to construct an accurate model for the shape of the Earth.

Let us now look at another classic example of the use of dimensional analysis of harmonic motion—Hooke's law. There are two ways to look at a harmonic oscillator. One is to consider all the forces acting on the mass, then use Newton's second law to derive a differential equation (an equation of motion) for the mass, solve it, and from the solution determine what happens if we, for example, change the mass. The second is to consider the dimensions of the quantities involved. We have a mass (M), we also have a spring with spring constant k, which has dimensions of force per unit length, or mass per unit time squared: $[k] = \frac{\text{Force}}{L} = \mathsf{MLT^{-2}/L} = \mathsf{MT^{-2}}$. For the frequency of oscillation, ω, we have $[\omega] = 1/\mathsf{T}$. Now we know that the

Table 6.3. Variables for Hooke's law.

Variable	Dimension	Description
k	MT^{-2}	Spring constant
m	M	Mass
x_0	L	Initial displacement of spring
t	T	Period of oscillation
A	L	Amplitude of motion
ω	T^{-1}	Frequency of vibration

frequency is a function of the spring constant and the mass, and that both sides of that equation must have the same sign. Since there is no mass in the dimension of the frequency, but it exists in the dimension of both the spring constant and the mass, we know that ω must depend on the ratio of k and m; that is, $\omega \sim k/m$. Now $[k/m] = T^{-2}$, and from $[\omega] = T^{-1}$; we conclude that we must have $\omega \sim \sqrt{k/m}$.

Imagine a mass (m) oscillating in one dimension on the end of a spring (with a spring constant k), and subject to no other forces. Besides m and k, the only other dimensional parameter is the initial displacement, x_0, of the spring from equilibrium. What can dimensional analysis tell us about the period (t), amplitude of motion (A) and total energy (E) of this system, in terms of these parameters? The dimensions of the various quantities in question are given in table 6.3.

As usual, we assume a power-law dependence and write $t = C\, m^\alpha k^\beta x_0^\gamma$, where α, β and γ are arbitrary powers to be determined, and C is a dimensionless constant. Taking the dimensions of both sides, we have

$$[t] = C[m]^\alpha[k]^\beta[x0]^\gamma \rightarrow T = M^\alpha(MT^{-2})^\beta L^\gamma = M^{\alpha+\beta}T^{-2\beta}L^\gamma, \tag{6.40}$$

as C is dimensionless. The solutions are: $\alpha = ½$, $\beta = -½$ and $\gamma = 0$. This gives the result that $t \propto \sqrt{m/k}$. Recall that the exact solution of the appropriate equation of motion yields $t = 2\pi\sqrt{m/k}$.

In this straightforward example, dimensional analysis alone determines how the result depends on the physical parameters, along with an undetermined dimensionless constant of order unity. If instead we ask how the amplitude A of the motion could depend on m, k and x_0, we find $A \propto m^\circ k^\circ x_0^1$. If we try the same approach for the total energy, we write $E = C_E\, m^\alpha\, k^\beta\, x_0^\gamma$, and matching dimensions, we find that $\alpha = 0$, $\beta = 1$ and $\gamma = 2$, so that $E \propto kx_0^2$, which is also correct, with $C_E = ½$. If instead of an initial x_0, we give the mass an initial velocity v_0, the analysis proceeds as above, but gives $t \propto \sqrt{m/k}$, but now $A \propto v_0\sqrt{m/k}$ and $E \propto mv_0^2$, again with the number of dimensional parameters matching exactly the number of dimensional constraints.

However, considering the more general problem with both an initial displacement and an initial velocity, we find that the system is under-constrained:

$$E \propto (m)^\alpha(k)^\beta(x0)^\gamma(v0)^\delta \rightarrow ML^2T^{-2} = M^\alpha(MT^{-2})^\beta L^\gamma(LT^{-1})^\delta = M^{\alpha+\beta}L^{\gamma+\delta}T^{-2\beta-\delta}. \tag{6.41}$$

We may solve for three of the exponents in terms of the fourth. Write: $\alpha = 1-\beta$, $\delta = 2-2\beta$ and $\gamma = 2\beta$; the energy of the oscillator can be written in the form:

$$E = C\ m^{1-\beta}k^{\beta}x_0^{2\beta}v_0^{2-2\beta} = C\ mv_0^2(kx_0^2/mv_0^2)^{\beta} = Cmv_0^2\Pi^{\beta}, \tag{6.42}$$

where the analysis of the problem has suggested a possible dimensionless ratio (energy, $kx_0{}^2$/energy, mv_0^2),

$$\Pi \equiv (kx_0^2/mv_0^2). \tag{6.43}$$

This new result reduces to the earlier simpler cases; for $\beta = 1$, we have the displacement-only case of $E \propto kx_0{}^2$, and for $\beta = 0$ we have $E \propto mv_0^2$.

Since the dependence in equation (6.42) is possible for any value of β, and any integral value of β is allowed; along with an arbitrary dimensionless constant C_β, we may write more generally that,

$$\begin{aligned} E &= (mv_0^2)(C_0\Pi_0 + C_1\Pi_1 + \dots) \\ &= (mv_0^2)\sum_{\beta} C_\beta\Pi_\beta = (mv_0^2)F(\Pi), \end{aligned} \tag{6.44}$$

so that any function of the dimensionless ratio Π is allowed. In this case, the appropriate function is straightforward, since,

$$\begin{aligned} E &= \tfrac{1}{2}mv_0^2 + \tfrac{1}{2}kx_0^2 \\ &= \tfrac{1}{2}mv_0^2(1 + kx_0^2/mv_0^2) \rightarrow F(\Pi) = \tfrac{1}{2}(1 + \Pi). \end{aligned} \tag{6.45}$$

This example is a model for the more general results established by Buckingham [1], who was one of the first to systematically describe methods of dimensional analysis, leading to the so-called Buckingham Π-theorem.

One might ask if gravity is involved in this model of harmonic oscillations. The answer is no, as we can also quickly see from dimensional analysis. The force of gravity is given by mg, because $[g] = \mathsf{LT}^{-2}$. Now if the frequency were to depend on g, there would have to be another factor to cancel the dependence on the length, as the frequency itself is length-independent. Neither m nor k has a length-dependence in its dimension, and so they cannot cancel the L in the dimension of g; the frequency therefore also cannot depend on g. The reason a simple pendulum's swing over small amplitudes has no dependence on mass is because the mass of the bob enters the dynamics in two different terms. There is the free-fall motion under gravity, and mass is also involved in inertia; that is, in Newton's second law. That means the resistance to changes in motion is directly proportional to the mass. However, the weight (a force = mg) on an object is also proportional to mass. Since the mass factors into both, the cause of changing motion and the resistance to changing motion, it cancels. For a harmonic mass-spring system, the mass still affects the inertia, but it does not cause, or force the harmonic motion. The spring (and its spring constant) is fully responsible for this force. So, mass only impacts the resistance to acceleration, and you notice that the more massive the object the slower it oscillates.

Reference

[1] Buckingham E 1914 On physically similar systems: illustrations of the use of dimensional equations *Phys. Rev.* **4** 345–76

IOP Publishing

Dimensional Analysis
The great principle of similitude
Jeffrey H Williams

Chapter 7

Rules of thumb, intuitive planning and physical insight

Where did we get that [Schrödinger's equation] from? It's not possible to derive it from anything you know. It came out of the mind of Schrödinger.
The Feynman Lectures on Physics

Dimensional analysis can sometimes seem to have a charmed quality. True dimensional analysis can allow someone without an in-depth knowledge of an area of physics to derive quite complex equations, but all the numerical factors are stripped away. This can be a hindrance and a danger, as well as a short-cut to the final answer. Someone who has experience of a field of physics, will have an innate feel for the magnitude of, and relation between the quantities they use. Someone without that experience may fall into a deep rabbit-hole using dimensional analysis. The loss of the numerical factors is a particular liability; however, many can be re-inserted into the final expressions from an in-depth feeling for the underlying physics. Of course, this *post-hoc* re-analysis of the problem raises the question of why bother to do the dimensional analysis, if you know the underlying physics?

One needs to take care and attention when using dimensional analysis. In this chapter, we will look at some of the rules of thumb that make dimensional analysis a rigorous means of solving problems. But it should be realised that there is no easy way to comprehend complex physics. We will look at some rules for generating and manipulating dimensional and nondimensional groups of variables; there will consequently be some overlap. These rules are provided to assist in selecting the variables needed to best define the quantity of interest; that is, how best to hypothesise a function of the dimensional variable form:

$$r = f(a, b, c, d, e \ldots). \tag{7.1}$$

doi:10.1088/978-0-7503-3655-0ch7

A function that is used in dimensional analysis to derive a final function, F, in terms of relevant nondimensional groups of variables,

$$R = F(A, B, C\ldots). \tag{7.2}$$

By using physical intuition (insight, and knowledge of physics—and not forgetting experimental data) and various rules of thumb, the list A, B, C … can be shortened from the initial list a, b, c, d, e …, which will facilitate and expedite a solution. But it must also be remembered that the final arbiter in any theoretical analysis is that something you have calculated can be compared to something that is measured, or is measurable.

The process of selecting variables for a dimensional analysis requires some experience, which is usually acquired by trial and error. If you include too many variables, you waste time trying to solve the problem. Saying that, however, if you miss the most important variables, then the analysis leads nowhere. This will generate incorrect, or inappropriate physical relations (this is where having experience is essential, as those rabbit-holes can be very deep), and a comparison with experimental results could then lead to further confusion. When comparing a new theoretical model to experimental data, it is important the model is able to 'collapse' the data onto as simple a scaling law as possible. If there are many obvious sub-sets within the supposedly collapsed data then this is an indication that something may be wrong, or missing from the initial hypothesis.

7.1 Dimensional variables

1. If a variable is known to appear in the problem under investigation exclusively within a certain algebraic combination, then you may hypothesise that combination as an independent variable. Consider some function of dimensional parameters, $r = f(x, y, z)$. If, for example, y and z have the same dimension, and you have good reason to believe they will occur only as a difference and never independently, then you might consider modifying the hypothesis to $r = g(x, y-z)$. An example would be two temperatures; instead of analysing T_1 and T_2 independently, use $\Delta T = T_1 - T_2$ (see section 7.4). If you suspect that the formulation will involve both differences, and the absolute values of y and z, then write, $r = f(x, y, y-z)$, and manipulate the dimensional analysis such that the difference, $y-z$, appears only in non-dimensional terms where it is important, and the parameter y appears in terms where it is important.

   Consider the dimensional analysis of the vertical linear spring via Hooke's law, with force constant, k; where $[k]$ = Force per length= $\mathsf{MLT^{-2}/L}$ (see section 6.3). The spring is compressed by a force of two masses, m_a and m_b. We know from the properties of harmonic oscillators that the relevant dimension will be a displacement, x, relative to the initial location of the spring at x_0. We might express the hypothesised function to be $x = f(x_0, m_a, m_b, k, g)$, where g is the acceleration due to gravity. To reduce the number of variables, we might analyse the following function $(x-x_0) = f(m_a + m_b, k, g)$.

Again, if you are convinced (perhaps by an initial experiment) that the variable will appear only as the product $k(x-x_0)$, then one could analyse $k(x-x_0) = f(m_a + m_b, g)$.

2. To decide whether or not to exclude a variable, it is often useful to consider if you can express it as a function of the remaining variables. Consider the following hypothetical function of dependent, dimensional variable, $r = f(x, y, z)$. If by physical insight, or by experimental observation, you can argue that parameter z is itself a function of x and y; that is, $z = g(x, y)$, then we may exclude z from the analysis and consider only $r = f(x, y)$. This is equivalent to considering nested functions; that is, consider some function $h = f_1(x, g)$, if we can argue that the variable g is $g = f_2(x)$, then we may write $h = f_3(x)$.
3. In hypothesising independent variables, you must include and retain physical constants that have units (for example, c = 299 792 458 m s^{-1}, but not π = 3.141 59...). Consider the equation relating energy (E) and mass (m) via the speed of light, $E = mc^2$; this may be written as $E = f(m, c)$. However, even though the speed of light is a constant of Nature, whose value has been fixed by the definition of the metre in the SI, we may not write $E = f(m)$. The dimensional constant is fundamental to the relationship. Energy (the dimensions of which are $\mathsf{ML^2T^{-2}}$) cannot be formulated without both dimensions of time and distance. This is the case for all constants with physical dimensions; for example, Newton's gravitation constant G, the Planck constant h and the Boltzmann constant k_B. On the other hand, we are free to exclude constants that are dimensionless. These dimensionless constants may be absorbed into the definition of the function you are considering. Consider the equation for the area of a circle, A, in terms of its radius, r; $A = 2\pi r$. We can write this equation, without loss of generality, as $A = f(r)$, since both π and the square of r are part of the meaning of f. This is physical intuition. In addition, we start with a hypothesised relationship involving a proportional relationship. To turn this proportional relationship into a mathematical equality requires a constant of proportionality, which will contain all the numerical constants.
4. Dimensional analysis involves manipulating combinations of variables via their base dimensions; the base dimensions are given in table 1.1, and correspond to the base quantities of the SI (see table 1.3). For the majority of problems in dimensional analysis, one is considering the base dimensions of mass, time, length, temperature and charge. When constructing your hypothesised function to describe the phenomenon of interest, if one or more of the dimensions appears in just one of the variables, and not in any other, then you may exclude that variable—unless you are using the Buckingham Π-theorem, when there are rules of distinguishing between dependent and free, or repeating variables (see chapter 5).

   Consider the following hypothesised function involving acceleration (a), velocity (x), length (y) and mass density (z): a ($\mathsf{LT^{-2}}$) = $f(x$ ($\mathsf{LT^{-1}}$), y (L), z ($\mathsf{ML^{-3}}$)). Looking at such a hypothetical function, one would need to look again at the inclusion of the mass density—it is the only variable involving mass.

Including this variable in the analysis will lead to problems, because z cannot be turned into a dimensional group, except by dividing by itself. Such a division yields unity, a dimensionless constant that can be (by the rule above) absorbed into the function definition. Hence, we would conclude that a must be described by $a = f(x, y)$.

5. If we are invoking geometric similarity, or geometric similitude (modelling of boats, or flying insects and airplanes, or estimating the running speed of a dinosaur—see later chapters), then all the characteristics of that shape are described by a single length scale. Thus, is it unnecessary to include other purely geometric descriptions? Consider some variable, r, as a function of a shape length, s, area, A, and volume, V, as in $r = f(s, A, V)$.

   We should instead write $r = f(s)$, $r = f(A)$, or $r = f(V)$, since all aspects of the shape are derivable from either s, A, or V. Any area could be expressed as $A = cs^2$, where c is a dimensionless constant for the selected shape. Then by the rules of thumb given here, we may exclude A if we include s, and vice versa. Similarly, including both density, ρ, and geometric length, s, precludes the inclusion of mass, m, since $m = K\rho s^3$, where K is a dimensionless constant, which become superfluous under geometric similarity.

   One important note on geometric similarity, particularly of relevance to problems in fluid mechanics, is the effect of surface roughness. The roughness element length scale, ε; that is, the size of irregularities that characterise the internal surface of pipes or conduits, or the bumps on the surface of a supposed sphere can introduce a second, important length scale in the characterisation of the geometry of the system under discussion. This inclusion is required, because experiments have revealed the significant influence of surface roughness on the behaviour of fluids in pipes. This second length variable in the dimensional analysis results in an additional dimensionless quantity known as the relative surface roughness (ε/d), where d is the macroscopic length scale used to characterise the geometry of the shape under discussion (see section 7.3).

7.2 Nondimensional variables

1. A function, for example, $Y = F(X)$ is a binary relation between two sets that associates every element of the first set to exactly one element of the second set. Functions were originally the idealization of how a varying quantity depends on another quantity; for example, the position of a planet as a function of time. Historically, the concept of what constituted a function was elaborated with the invention of calculus. Thereafter, functions were defined as being differentiable entities—containing a high degree of symmetry. A function is also a process, or a relation that associates each element x of a set X, the domain of the function, to a single element y of another set Y (possibly the same set), the codomain of the function. It is customarily denoted by letters such as f, g and h. If the function is labelled f, this relation is denoted by $y = f(x)$, where the element x is the argument or input of the

function, and y is the value of the function, the output. A function is uniquely represented by the set of all pairs $(x, f(x))$, called the graph of the function. When the domain and the codomain are sets of real numbers, each such pair may be thought of as the Cartesian coordinates of a point in the plane. The set of these points is called the graph of the function; and is the usual manner of illustrating the function.

The generality of a relation such as $Y = F(X)$, allows us to manipulate this expression and generate new functions that relate the relevant variables. This is particularly useful for functions of nondimensional variables; for example, we may wish to relate a function between the square of Y and the inverse of X, for:

- Ensuring that the dimensional analysis generates a recognizable nondimensional parameter such as a Reynolds number, or a drag coefficient.
- Isolating weak dependence in the constituting variables.
- Inverting a function or a constant.
- Isolating a variable such that it appears in only one nondimensional group.

To manipulate a group of nondimensional parameters, we recognize that each nondimensional quantity can be multiplied by itself, or by another such parameter raised to a non-zero power, and the result is still a nondimensional quantity (see section 4.1). Consider some function relating the nondimensional dependent variable H to nondimensional parameters, X, Y and Z: $H = F_1(X, Y, Z)$. H^a, X^b, Y^c and Z^d are each nondimensional for non-zero real values of a, b, c and d. Similarly, the product $H^a X^b Y^c Z^d$ is also nondimensional for such exponents. Hence, we may write any of the following:

$$H^a = F_2(XY^b, Y, Z), \tag{7.3}$$

$$H = F_3(X, YZ^b, Z), \text{ or} \tag{7.4}$$

$$HX^a = F_4(X, YZ, X^aY^bZ^c), \tag{7.5}$$

as three of the many examples for non-zero values of the exponents. The exponents, a, b, c and d are arbitrary, real and non-zero powers of the nondimensional parameter—and may be positive or negative. In manipulating such quantities, we are preserving the number of dimensions of the function. Hence, we cannot simply eliminate a nondimensional quantity by dividing by itself.

Dimensional analysis sometimes generates an expression such as: $F(X) = 1$, where a function of a nondimensional variable X is equal to unity. This is rearranged by inverting it to give $X =$ constant. That is, if a function of a variable is always equal to a constant (or a number), then the variable itself must be a constant.

2. Isolating dependences: If the form of the function is known for one of the independent variables x, while holding the others constant, then this

dependence may be evaluated explicitly. This form of evaluation is most successful if the function of nondimensional variables is manipulated to have only a single nondimensional term, X, which contains the variable in question, x. The known dependence may arise from:

- Experiment, where the independent variable is varied while other parameters are held constant.
- A strong analogy with a similar, known problem.
- The application of physical insight when defining the hypothesised function.

The most commonly hypothesised known dependence for an independent variable, x, is a power-law of the form x^a. Consider the function of dimensional parameters $r = f(a, b, c, d, e)$. Now consider a hypothetical case such that r is known to be directly proportional to b raised to the sth power, as in $r \propto b^s$, in the case where the other variables are held constant. If the corresponding function of the nondimensional variables is, for example,

$$R = F(A = a/b,\ B = ba/c^2,\ C = d/e), \tag{7.6}$$

then we have not isolated the variable in question, b, and we cannot use this particular rule of thumb. If, however, we manipulated the function as in the rule above to derive,

$$R = G(a/b,\ a^2/c^2,\ d/e), \tag{7.7}$$

then the dependence on b is confined to the first term within the parenthesis. If the other variables do not depend upon b, then we may hypothesise that

$$R = \left(\frac{a}{b}\right)^{-s} H(a^2/c^2,\ d/e). \tag{7.8}$$

That is, we explicitly write the dependence (proportionality) for the only nondimensional parameter that contains the dimensional variable of known dependence. This dependence is formulated as a pre-factor multiplying the unknown function. This respects the principle of dimensional homogeneity, since both the dependent variable (on the LHS) and the pre-factor are dimensionless. The important thing is to isolate the quantity of known dependence (b in the above example). Given this isolation, the nondimensional parameter that contains the variable of interest may also contain other variables (here a).

One of the best examples of this rule of thumb is to be found in the founding publication of dimensional analysis [1]. Consider a function describing the total heat transfer, Q, (that is, power, or energy per time, $\frac{\partial E}{\partial T}$) from a thin filament immersed in a fluid of specific heat C, density ρ, thermal conductivity k, moving at a velocity v. Lord Rayleigh hypothesises the following function to describe this model,

$$Q = f(a,\ L,\ \Delta T = T - T_{\infty},\ v,\ \rho C,\ k). \tag{7.9}$$

Here, a and L are the radius and length of the heating filament, respectively. As given in these rules, Lord Rayleigh has already reasoned that the problem will not depend upon two separated temperatures, but on a single temperature difference; neither will it depend upon a separate density variable and a separate variable for heat capacity, but on a combined variable, which will have dimensions $[\rho C] = [\rho][C]$. Thus, a nine variable problem has become a problem of seven variables. A dimensional analysis using the Buckingham Π-theorem yields the following for the hypothesised function (see section 7.4):

$$\frac{Q}{kL\Delta T} = F\left(\frac{av\rho C}{k}, \frac{L}{a}\right). \tag{7.10}$$

Next, Lord Rayleigh considers the idea that the dimensional heat transfer rate, Q, will scale proportionately as some power of the velocity, v. Lord Rayleigh attributed this insight to the French engineer Joseph Valentin Boussinesq (1842–1929), who found a solution in the limit of thin thermal boundary layers and inviscid flow, when the problem becomes one of one-dimensional heat transfer. In a simpler case, the heat transfer rate scales with the velocity as $Q \propto \sqrt{v}$.

In applying this rule of thumb, we first ensure that the dependent variable in question is confined to a single nondimensional group, $\frac{av\rho C}{k}$. Second, we note that the other variables in this nondimensional group, a, C and k, do not depend upon the identified variable of interest, v. Then we infer that the dependence will apply to the entire nondimensional group involving the velocity. We evaluate the function as

$$\frac{Q}{kL\Delta T} = \sqrt{\left(\frac{av\rho C}{k}\right)}G\left(\frac{L}{a}\right). \tag{7.11}$$

These ideas can be useful in simplifying functions of dimensional variables. The test of the veracity, or otherwise, of the full dimensional analysis will be revealed if the final model equation is able to rationalise, or collapse experimental data.

3. If the function of interest is a weak function of one parameter, seek to make that weak parameter appear as rarely as possible. Consider the drag force on a sphere of diameter, d, travelling at a velocity, v. For high velocities, we may hypothesise the drag force to be $F_d = f(d, v, \rho, \mu, c)$. Here, ρ, μ and c are the mass density of the fluid, the dynamic viscosity of the fluid and the local speed of sound of the fluid (a measure of the fluid's compressibility). A first application of dimensional analysis would yield the following function (there are six variables composed of three base dimensions),

$$F_d/\rho c^2 d^2 = F_1\left(\frac{\rho c d}{\mu}, \frac{v}{c}\right), \tag{7.12}$$

where v/c is a Mach number, Ma. Although dimensionless, the variables have been complicated by the presence of the speed of sound, and we have lost the standard definition of drag coefficient and Reynolds number. If we believe the phenomenon will be a weak function of Mach number (Ma < 0.3), we should isolate the weak functionality. This may be remedied by:

$$F_d/\rho c^2 d^2\left(\frac{v}{c}\right)^{-2} = F_2\left(\frac{\rho c d}{\mu}, \frac{v}{c}, \frac{v}{c}\right) \tag{7.13}$$

so

$$F_d/\rho v^2 d^2 = F_3\left(\frac{\rho v d}{\mu}, \frac{v}{c}\right). \tag{7.14}$$

The weak dependence of the speed of sound is now localised in the final term. It is a matter of physical insight to decide whether or not a particular variable is weakly important or not. For the present case of the speed of sound, it is knowledge of the physics going on in the phenomenon (perhaps derived from experiment) that should suggest whether or not the drag is sensitive, or insensitive, to the Mach number. In which case, we would write,

$$F_d/\rho v^2 d^2 = F_3\left(\frac{\rho v d}{\mu}\right). \tag{7.15}$$

4. After having completed a dimensional analysis, you may subsequently convince yourself that one of the nondimensional parameters can be expressed as a function of the others. Or you may have just acquired new experimental evidence confirming such a dependence. Consider the following function of nondimensional parameters, A, B, C and D: $A = F(B, C, D)$. If you can define some known, or unknown function, such that $B = G(C, D)$, or you can confirm this relation by experiment, then you can simplify the hypothesised function to $A = G(C, D)$.
5. As mentioned, we may always absorb nondimensional parameters that are constant into a function. The constant just becomes part of the definition of the function. In problems involving geometric similarity, we implicitly absorb all of the dimensionless and definable length, area and volume ratios of the geometry into the function. In some cases, it may be possible to apply this approach in an approximate fashion. Hence, we can simplify a function of nondimensional variables by absorbing one whose variability has only a negligible effect upon the dependent variable. In this manner, we treat the nondimensional variable as approximately constant. Consider the function $A = F(B, C, D)$. If variation of the nondimensional parameter, D (over the whole range of expected values of D), has a negligible effect on A (compared to the effect of B and C), then consider treating this nondimensional variable as if it were a constant and so absorb it into the function, and the associated hypothesis; that is, $A \cong F(B, C)$. The notation says that A and $F(B, C)$ are

congruent (equivalence of geometric shapes and size $\Delta ABC \cong \Delta XYZ$) to each other. Such an insight may arise from experiment, or by physical insight.

6. It is never a good idea to merely eliminate a variable merely because it is small. Such elimination should only be made with knowledge of the function, or with some additional source of information (experiment, or physical insight). Consider the following hypothetical function in terms of nondimensional variables A, B, C and D: $A = F(B, C, D)$. Consider a limit such that D approaches a small value. This limit is not sufficient to determine the strength, or weakness of the function's dependence; for example, consider that the underlying function may have the form $A \approx CD + BD$. Hence, the effect of D on A is strong in the limit of vanishing D. Also, if the limiting behaviour also affects the magnitude of C or B, then we must evaluate the product CD versus the product BD in such a limit. Physical insight and experiment are required to eliminate variables.
7. Finally, and most importantly: The rule of relevance. In a dimensional analysis, only include relevant physics; for example, when considering gravity, use g for a rock, or an apple on the surface of the Earth, but G for the interaction of the Sun, moon and Earth.

7.3 Eliminating a variable you suspect could be negligible: the design of golf-balls

Let us now look at a system where even a slight change in geometry can have a significant, one might say disproportionate, effect on the subject's dynamics, thereby making it difficult to approximate the behaviour of one body of given experimental data on the behaviour of another related subject—a breakdown of the principle of similitude. In chapter 8, we look at the effect of a drag force on the dynamics of bodies. We discuss these phenomena in terms of drag coefficients on similar smooth spheres. But what happens when the surface of the sphere is not smooth? Roughened to even a slight extent (0.1% relative roughness compared to the diameter of the sphere), the surface dynamics of the fluid at the surface (the boundary layer) will perturb the dynamics of the entire sphere sufficiently to make a comparison with a truly smooth sphere meaningless.

To analyse this problem of the dynamics of a sphere in a moving fluid, or a sphere falling or rising in a static fluid, by dimensional analysis requires the inclusion into the hypothesised function for an addition length scale—the surface roughness, we are already considering the diameter of the sphere. Thus, using the variables in table 7.1, we set up a model for the magnitude of the drag force on a sphere with a roughened surface as:

$$F_d = f_1(v, d, \varepsilon, \rho, \mu). \tag{7.16}$$

Using dimensional analysis, we are able to derive (one can readily see in table 7.1 the origin of one of the three nondimensional groups that define this problem; that is, ε/d),

Table 7.1. Variables for the dynamics of a roughened sphere.

Variable	Dimension	Description
F_d	MLT^{-2}	Drag force on sphere
v	LT^{-1}	External flow velocity
d	L	Diameter of sphere
ε	L	Roughness element
ρ	ML^{-3}	Density of fluid
μ	$ML^{-1}T^{-1}$	Dynamic viscosity of fluid

$$F_d/\rho d^2 v^2 = G\left(\frac{\mu}{\rho v d}, \frac{\varepsilon}{d}\right). \tag{7.17}$$

This analysis is analogous to that discussed in flow in conduits, see section 5.2.3. Invoking the rules of thumb for nondimensional variables, we may reformulate this in terms of a more conventional nomenclature as:

$$C_d = F\left(\text{Re}, \frac{\varepsilon}{d}\right), \tag{7.18}$$

where $\text{Re} \equiv \frac{\rho v d}{\mu}$ is the Reynolds number and $C_d = F_d/\frac{1}{2}\rho v^2\ (\pi d^2/4)$ is the drag coefficient of a roughened sphere of diameter d.

The literature is full of experimental data for the drag of both rough and smooth spheres in terms of the nondimensional parameters of the drag coefficient and Reynolds number [2, 3] with results reproduced in [4]. It is seen that spheres with turbulent boundary layers (roughened surfaces) have smaller wakes than smooth spheres, and so experience significantly less drag. As an example of this small geometric change, which generates a significant variation in the drag forces experienced by a sphere, let us look at a golf ball. Engineers and scientists in the golf industry study the impact between a golf-club and a golf ball to determine the launch conditions. This impact lasts, typically, only 1/2000 of a second, but it establishes the ball's velocity, launch angle and spin rate. After this brief impact, the ball's trajectory is controlled entirely by gravity and aerodynamics—no matter how much the golfer hopes or curses. As a result, aerodynamic optimization—achieved through a dimple pattern design—is a critical part of overall golf ball development (see figure 7.1).

Research reveals that a smooth golf ball, hit by a professional golfer will travel only about half as far as a dimpled golf ball. Most golf-balls have between 300 and 500 dimples, which have an average depth of about 0.25 mm (figure 7.1). The lift and drag forces on a golf ball are very sensitive to dimple geometry: a depth change of 10% will change the ball's trajectory and the distance achieved.

Air exerts a force on any object moving through it. This force has two components: lift and drag. Drag acts to directly oppose motion, whereas lift acts in a direction perpendicular to motion. A moving object has a high-pressure area on

Figure 7.1. A dimpled golf ball (This 'teed up golf ball on Plastic tee' image has been obtained by the author from the Wikimedia website where it was made available under a CC BY-SA 4.0 license. It is included within this article on that basis. It is attributed to JManning11.)

its front-side. Air flows smoothly over the contours of the front-side and eventually separates from the object toward the reverse-side. A moving object also leaves behind a turbulent wake region where the airflow is agitated, or perturbed. This disturbance results in a pressure difference between the two sides of the object—with lower pressure behind it. The size of the wake affects the amount of drag on the object. Dimples on a golf ball create a thin turbulent boundary layer of air that tends to cling to the ball's surface. This allows the smoothly flowing air to follow the ball's surface a little farther around the reverse-side of the ball, thereby decreasing the size of the wake. Experiments show that a dimpled ball has about half the drag of a smooth ball. Dimples also affect lift. A smooth ball with backspin creates lift by channelling the airflow such that the golf ball acts like an airplane's wing. The spinning action makes the air pressure on the bottom of the ball higher than the air pressure on the top; this imbalance creates an upward force on the ball. Ball spin contributes about one half of a golf ball's lift. The other half is provided by the dimples, which allow for optimization of the lift force.

There is a large amount of literature on the subject of the best design of golf ball. This research seeks to measure the drag coefficient, C_d, as a function of Reynolds number for golf-balls of different design [5]. Figure 8.4 shows the general behaviour of C_d versus a vast range of Re; golf-scientists are interested in C_d over the range of order Re = 1–10 $\times 10^4$ [5]. That is, region 4 in figure 8.4.

Results of such studies demonstrate that the drag coefficient of a golf ball varies significantly due to varied dimple geometry. The results indicate that the increase of the dimple depth, or surface roughness of the golf ball can shift the transition to a lower Reynolds number and increase the drag coefficient in the transcritical regime. The results also establish a positive linear correlation between relative roughness and drag coefficient. Consequently, be careful when excluding variables from your dimensional analysis, just because you imagine they are only involved in a weak dependence upon the dependent variable.

Table 7.2. Variables used in Rayleigh's paper on dimensional analysis [1].

Variable	Definition	Modern dimensionality	Rayleigh's dimensionality
Q	Heat transfer rate (energy per unit time)	$\mathrm{ML^2T^{-2}/T = ML^2T^{-3}}$(Power)	$(\mathrm{ET^{-1}}) \equiv$ Power in the SI $(\mathrm{ML^2T^{-2}/T = ML^2T^{-3}})$
a	Radius of heating filament	L	L
L	Length of heating filament	L	L
T–T_∞	Cylinder to fluid temperature difference	θ	θ
ρc_p	Specific heat of the fluid per volume	$\mathrm{ML^{-1}\,T^{-2}\,\theta^{-1}\,(ML^{-3}.\,L^2T^{-2}\,\theta^{-1} = ML^{-1}T^{-2}\,\theta^{-1})}$	$\mathrm{EL^{-3}\,\theta^{-1}}$
V	Velocity of fluid	$\mathrm{LT^{-1}}$	$\mathrm{LT^{-1}}$
k	Thermal conductivity of fluid	$\mathrm{MLT^{-3}\,\theta^{-1}}$	$\mathrm{EL^{-1}\,T^{-1}\,\theta^{-1}}$

7.4 The Rayleigh–Riabouchinsky Paradox

The problem Lord Rayleigh dealt with in most detail in the founding publication of dimensional analysis [1] concerned the modelling of the steady-state heat transfer rate, Q, from a long, thin cylindrical filament of length, L, and radius, a. The fluid was deemed inviscid (the viscosity of the fluid is assumed to be zero); the variables Rayleigh deemed to be essential for modelling this problem are listed in table 7.2. Also listed in table 7.2 are the modern (SI) dimensions of the same quantities. See the analysis involving equations (7.9)–(7.11) above.

Note that Rayleigh's dimensions are different from those we would give the same variables today. Instead of M, L, T and θ (for temperature), Rayleigh used E (energy), L, T and θ. He did not use an independent dimension of mass, but incorporated it into his dimension for energy, E. This is not surprising, as at that time the physics community was divided as to the existence of atoms and molecules [6], and it was only the chemistry community that was interested in individual molecules possessing mass—the physics community used continuum models.

Lord Rayleigh's approach was, as in many steady-state heat transfer problems, to consider the difference in temperature as a single variable, T–$T_\infty = \Delta T$. In addition, Rayleigh expects the problem to scale not independently with both fluid density (ρ) and specific heat (c_p) of the fluid, but only as the product of these two physical properties; the SI unit for specific heat is joule per kelvin per kilogram, and density is kilogram per metre cubed ($\mathrm{ML^2T^{-2}\,M^{-1}\,\theta^{-1} = L^2T^{-2}\,\theta^{-1}}$, and density = $\mathrm{ML^{-3}} \rightarrow \mathrm{ML^{-1}T^{-2}\,\theta^{-1}}$). Rayleigh would therefore have assumed a model for his problem as

$$Q = f(a, \Delta T, V, \rho c_p, k). \tag{7.19}$$

Setting up the dimensional equation to evaluate the exponents gives (we will use Rayleigh's dimensions)

$$ET^{-1} = (L)^x(\theta)^y(LT^{-1})^z(EL^{-3}\theta^{-1})^u(EL^{-1}T^{-1}\theta^{-1})^v = E^{u+v}\theta^{y-u-v}T^{-z-v}L^{x+z-3u-v}. \quad (7.20)$$

There are five unknown quantities, and only four equations; so, Lord Rayleigh was only able to state:

$$Q = ka\Delta T\left(\frac{aV\rho cp}{k}\right)^z. \quad (7.21)$$

Since z is undetermined, any number of terms of this form may be combined, and all that we can conclude is that

$$Q = ka\Delta TF\left(\frac{aV\rho cp}{k}\right), \quad (7.22)$$

where F is an arbitrary function of the variable $\frac{aV\rho cp}{k}$.

Lord Rayleigh then demonstrated how kinematic viscosity (with dimensions L^2T^{-1}) could be added as an additional variable. As there are no additional base dimensions involved in the addition of kinematic viscosity (υ) as a variable, we now have six unknown quantities for four equations, and so there will be two undetermined or arbitrary exponents, and Rayleigh was able to write:

$$Q = ka\Delta T\left(\frac{aV\rho cp}{k}\right)^z\left(\frac{\upsilon}{k}\right)^w = ka\Delta TG\left(\frac{aV\rho cp}{k}, \frac{\upsilon}{k}\right), \quad (7.23)$$

where G is an arbitrary function of the two variables $\frac{aV\rho cp}{k}$ and $\frac{\upsilon}{k}$. The latter of these two being the ratio of the diffusivities for momentum and for temperature.

Four months after the publication of Rayleigh's short article (March 1915), the Russian (although he left Russia after the October Revolution, and lived subsequently in France) fluid dynamicist, Dimitri Pavlovitch Riabouchinsky (1882–1962), published a very short comment on Rayleigh's paper. Riabouchinsky was taking issue with Rayleigh's formulation of the variables defining the problem of interest. Although there are two variables, each with dimensions of L mentioned in table 7.2; that is, a and L, Rayleigh formulated his expression for the heat transfer rate using only a single geometrical length, a (see the above equations). Riabouchinsky commented: '*Lord Rayleigh gives this formula* [$Q = ka\ \Delta T\ G(\frac{aV\rho cp}{k})$] *considering heat, temperature, length and time as four 'independent' dimensions. If we suppose that only three of these quantities are really independent, we obtain a different result. For example, if the temperature is defined as the mean kinetic energy of the molecules, the principle of similarity allows us to affirm that* [$Q = ka\ \Delta T\ G(V/ka^2),\ \rho c_p a^3$].' (The equations given in this quote use modern notation for clarity.)

Dimitri Riabouchinsky is raising the great question of that period—the nature of the structure of matter, and how—if you accept the reality of the existence of atoms and molecules—can you explain the origin of the central pillar of 19th Century science, the second law of thermodynamics? Riabouchinsky points out that we can express temperature in terms of the kinetic theory of molecules, apparently removing the independent dimension of temperature from the problem. However, the result of this change would be the generation of an additional nondimensional parameter. That is, additional physics (physical insight) seems to have resulted in a less powerful result, and less overall understanding of the phenomenon. This issue has been termed the Rayleigh–Riabouchinsky Paradox.

Indeed, this paradox puzzled Lord Rayleigh. Two weeks after the publication of Riabouchinsky's comment, Lord Rayleigh replied. '*It would indeed be a paradox if the further knowledge of the nature of heat afforded by the molecular theory put us in a worse position in dealing with a particular problem... It would be well worthy of discussion.*' There was further discussion and progress, but illumination came slowly. Essentially, the discussion between Rayleigh and Riabouchinsky, on the nature of heat and how we might represent this nature is related to discussions between Albert Einstein and Ernst Mach. The Rayleigh–Riabouchinsky discussion is a local example, from the field of dimensional analysis, in the much more general discussion about the reality of molecules [6, 7].

Let us look further at Riabouchinsky's comment. He proposed removing the base dimension of temperature (θ), and replacing it with a description of the average kinetic energy of the molecules. Thus, the dimension of $\Delta T = T - T_{\infty}$, would become $\mathsf{ML^2T^{-2}}$. Of course, when we look at this paradox today, we have the advantage of having been trained in the SI system of units. And it is in the SI that we find the means of relating the average energy content of a macroscopic sample of fluid to the energy content of the molecules that compose that large sample. The number of molecules in a sample is given by Avogadro's number ($N_A = 6.022\ 140\ 76 \times 10^{23}$ per mole), and to relate the temperature of this bulk to the energy content of the individual molecules we need to invoke the Boltzmann constant ($k_B = 1.380\ 649 \times 10^{-23}$ J K^{-1}). The Boltzmann constant is the proportionality factor that relates the average relative kinetic energy of particles in a fluid with the thermodynamic temperature of the fluid. It occurs in the definitions of the kelvin and the gas constant, R, and in Planck's law of black-body radiation, and in the Boltzmann entropy formula. The Boltzmann constant has the dimension of energy divided by temperature—the same as entropy. As part of the 2019 redefinition of SI base units, the Boltzmann constant is one of the seven defining constants that have been given exact definitions. The Boltzmann constant is defined to be exactly $1.380\ 649 \times 10^{-23}$ J K^{-1}. The product $N_A.k_B = R$, the gas constant (8.314 462 618 153 24 J K^{-1} mol^{-1}), and thus we are today able to relate the thermodynamic properties of material in bulk quantities to the dynamics and behaviour of the molecules that compose the bulk [8].

Of course, none of this was known back in 1915. Going back to what Rayleigh and Riabouchinsky knew, the relationship between energy and temperature was not $E \sim \Delta T$, but $E/k_B \sim \Delta T$. If we wished to re-do the dimensional analysis of the problem of heating a fluid, the variables in table 7.2 would need to be modified; in particular, ΔT would become $k_B \Delta T$ and ρc_p would become $\rho c_p / k_B$. A dimensional analysis of these variables leads to the following expression for the rate of heat transfer,

$$Q/ka\Delta T = G(k_B V/ka^2,\ \rho c_p a^3/k_B). \tag{7.24}$$

There is no advance in understanding the phenomenon—we are still a long way from a single formula to define the effect proposed by Lord Rayleigh. Riabouchinsky's choice of variables has introduced an unnecessary level of complexity into the problem. The introduction of the properties of the molecules forming the fluid being heated is irrelevant (to the problem in hand). The problem was sufficiently well-formulated using continuum models of the nature of matter. That is, in models where time and space averages lead directly to the desired concepts of continuum properties such as conductivity, heat transfer rates, density and specific heat. Picking out one variable and replacing it with another variable more suitable for describing the behaviour of individual molecules only confuses the situation. One is mixing continuum models of the structure of matter, with the molecular theory of the structure of matter. But an in-depth look at the Rayleigh–Riabouchinsky Paradox does allow us to formulate a general rule for dimensional analysis. Dimensional analysis should include only the relevant physics, and should seek to exclude inappropriate physics. Here, relevance is determined by the questions we are asking, and the scale of the forces involved.

The lesson here is an important one because it illustrates the role played by the fundamental constants. Consider the Planck constant, h; it would be completely inappropriate to introduce it into a problem of classical dynamics. For example, any solution of the scattering of two billiard balls will depend on macroscopic variables such as the masses, velocities, friction coefficients, and so on. Since billiard balls are made of protons, it might be tempting to include as a dependent variable the proton–proton total cross-section, which, of course, involves h. To an experienced physicist this would clearly be totally inappropriate, but this confusion is analogous to what Riabouchinsky did in Boussinesq's problem, as given by Lord Rayleigh [1].

References

[1] Rayleigh L 1915 The principle of similitude *Nature* **95** 66–7

[2] Achenbach E 1975 The effects of surface roughness and tunnel blockage on the flow past spheres; *J. Fluid Mech.* **65** 113–25

[3] Achenbach E 1972 Experiments on the flow past spheres at very high Reynolds number *J. Fluid Mech.* **54** 565–75

[4] Santiago J G 2019 *A First Course in Dimensional Analysis: Simplifying Complex Phenomena Using Physical Insight* (Cambridge, MA: MIT Press)

[5] Chowdhury H, Loganathan B, Wang Y, Mustary I and Alam F 2016 A study of dimple characteristics on golf ball drag *Procedia Eng.* **147** 87–91

[6] Williams J H 2018 *The Molecule as Meme* (San Rafael, CA: Morgan & Claypool)

[7] Pais A 1982 *Subtle is the Lord: The Science and the Life of Albert Einstein* 1982 (Oxford: Oxford University Press) ch 5

[8] Jeffrey H 2020 *Williams: Defining and Measuring Nature: The Make of All Things* 2nd edn (Bristol: IOP Publishing)

IOP Publishing

Dimensional Analysis
The great principle of similitude
Jeffrey H Williams

Chapter 8

Continuum forces

"I am an old man now, and when I die and go to heaven there are two matters on which I hope for enlightenment. One is quantum electrodynamics, and the other is the turbulent motion of fluids. And about the former I am rather optimistic".
Horace Lamb (1849–1934).
Address to the British Association for the Advancement of Science in 1932

We have seen how the foundations of dimensional analysis were laid in the first years of the last century; particularly with the work of Lord Rayleigh. Interestingly, Lord Rayleigh's paper in *Nature* of 1915 [1], which demonstrated the power of the technique for handling complex problems without the need for difficult experiments, or the solution of coupled differential equations, appeared at the same time as another application of dimensional analysis. When presenting his model of the hydrogen atom in 1913, Niels Bohr used an argument based on the dimensions of variables in the introduction of his article [2]. The classical mechanical problem of calculating the orbits of the electron around the proton in the hydrogen atom, involves only two input variables: the mass of the electron, and the constant determining the strength of the force between the two charged-particles. But these two input variables cannot be combined into a physical quantity with dimensions of length needed to explain the size of the atom. On the other hand, when the Planck constant is introduced as a third variable, dimensional analysis gives a characteristic length—the Bohr radius—that agrees with the known order of magnitude of the size of the atom (see chapter 12).

8.1 The basic concepts of fluid mechanics

It is in fluid mechanics that dimensional analysis is most often to be found. This is a huge and difficult subject, with a long history not only in the field of physics, but also

doi:10.1088/978-0-7503-3655-0ch8

in mathematics and engineering. A fluid is described as any substance that deforms continuously when subjected to a shear stress; that is, a force tending to cause deformation by slippage along a plane or planes parallel to the imposed stress. A fluid will deform continuously, regardless of the magnitude of the stress. It is important to note that both gases and liquids are classed as fluids. A solid will also deform under stress, but as long as the elastic limit of the solid is not exceeded it does not continue to deform when the stress is removed.

The central physical property related to the deformation of fluids is the viscosity. Isaac Newton defined viscosity as: '*the resistance arising from the want of lubricity in the parts of a fluid.*' It provides a measure of stresses associated with gradients in the velocity of the fluid. For Newtonian fluids, the shear stress, τ (force per unit area, so it is like pressure in a gas) within a fluid, or at the fluid/solid interface is directly proportional to the local velocity field gradient. In a velocity field varying in one dimension, for example, y, we may express this quantity as $\tau \propto \frac{dv}{dy}$; this is Newton's law of fluids, with v being the fluid velocity. The constant of proportionality, which turns this relationship into an equality, is called the dynamic viscosity, μ; thus, $\tau = \mu \frac{dv}{dy}$. The dimensions of this property are: $[\tau] = \mathsf{MLT^{-2}}.\ \mathsf{L^{-2}} = \mu \frac{dv}{dy} = \mu\ (\mathsf{LT^{-1}}.\ \mathsf{L^{-1}})$; and so $[\mu] = \mathsf{MLT^{-2}}.\ \mathsf{L^{-2}/LT^{-1}}.\ \mathsf{L^{-1}} = \mathsf{ML^{-1}T^{-1}}$ (kg $m^{-1}s^{-1}$ in the SI). The ratio of dynamic viscosity to the density of the fluid (ρ) often arises in the analysis of problems in fluid mechanics, and it is called the kinematic viscosity, v. Thus, $v = \mu/\rho$; which has the dimensions $\mathsf{ML^{-1}T^{-1}/ML^{-3}} = \mathsf{L^2T^{-1}}$ ($m^2\ s^{-1}$ in the SI). The kinematic viscosity is interpreted as quantifying the diffusion of momentum along the velocity gradient. The relationship with other mechanical problems can be seen by comparing the dimensions of viscosity with the dimensions of pressure, $\mathsf{MLT^{-2}/L^2} = \mathsf{ML^{-1}\ T^{-2}}$. Physically, the viscous force is relevant to determining the drag force on an object moving in a fluid, because there are two sets of forces opposing motion:

- Inertial forces—the object pushes the medium apart as it moves, and
- Viscous forces—the fluid in the immediate vicinity of the object is pulled along at the speed of the object while fluid far away from the object is at rest. Thus, the fluid experiences a differential flow.

The viscosity of a fluid is a measure of its resistance to deformation. If you apply a pressure (force by area) to a solid, it will deform slowly. If you apply a pressure to a fluid, it will deform rapidly by flowing. Viscosity can be conceptualized as quantifying the internal frictional force that arises between adjacent layers of fluid that are in relative motion. For instance, when a fluid is forced through a tube, it flows more quickly near the axis of the tube than near the walls. In such a case, experiments show that some stress (such as a pressure difference between the two ends of the tube) is needed to sustain the flow through the tube (see example in chapter 5). This is because a force is required to overcome the friction between the layers of the fluid, which are in relative motion: the strength of this force is proportional to the viscosity. Viscosity is the material property which relates the viscous stresses in a material to the rate of change of a deformation (the strain rate).

Although it applies to general flows, it is possible to visualize and define in a simple shearing flow, such as a planar Couette flow (see figure 8.1). Since the shearing flow is opposed by friction between adjacent layers of fluid (which are in relative motion), a force is required to sustain the motion of the upper plate. The relative strength of this force is a measure of the fluid's viscosity.

In such a system, the flux, or flow of momentum through some area, A, can be expressed in terms of two velocities of the fluid: the local velocity relative to the area (velocity that determines flow rate across the area), and the velocity with respect to some internal reference frame that determines the fluid's momentum, and appears in Newton's second law [3]. For a one-dimensional flow at velocity, v, we can write:

$$\text{flow of momentum} = \rho v^2 A = \text{a force (dimension} = \mathsf{ML^{-3}.\ L^2T^{-2}.\ L^2} = \mathsf{MLT^{-2}}). \tag{8.1}$$

Changes of this quantity indicate reaction forces associated with inertia. The reader will discover that the group of variables with dimensions equal to ρv^2 occurs in many areas of fluid mechanics; for example, in the Bernoulli equation to estimate pressure rise associated with the frictionless deceleration of a fluid stream [3]. This rise in pressure is given by:

$$\text{dynamic pressure} = \tfrac{1}{2}\rho v^2. \tag{8.2}$$

The force associated with this pressure rise is then a force equal to the change in momentum of that stream:

$$\text{maximum force imparted by dynamic pressure per unit area} = \tfrac{1}{2}\rho v^2 A. \tag{8.3}$$

There are two regimens of fluid flow: laminar motion and turbulent motion. In laminar motion, the fluid is seen to move in laminae, or layers, and the stream is said to be steady and smooth. Such flow is characterised by a balance between viscous and pressure forces. A laminar flow will cease if you remove the applied pressure.

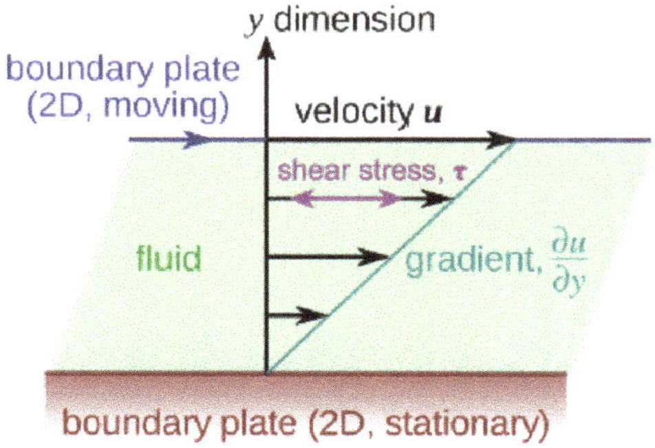

Figure 8.1. Simple Couette configuration between two infinite flat plates; demonstrating laminar shear in a fluid. This 'laminar shear' image has been obtained by the author from the Wikimedia website where it was made available under a CC BY-SA 4.0 license. It is included within this article on that basis. It is attributed to Duk.

The fluid comes to a halt due to the viscosity force. Turbulent flow is characterised by a high ratio of inertial forces to viscous forces. Turbulent flows are chaotic, fluctuating, and have a complex 3-D structure. A flow is turbulent if, after removing the forcing function, the fluid continues to deform in complex unsteady motions.

Whether or not a fluid flow will be laminar or turbulent is related to its Reynolds number, Re, where $\mathrm{Re} \equiv \frac{\rho v L}{\mu}$ (see section 5.2.2). Here, v and L are the characteristic velocity and length-scale of the flow, and a low value of Re implies a laminar flow, and a high value of Re implies inertial effects, and if Re is sufficiently high, a turbulent flow. The transition between the two flow regimens depends upon the flow geometry (jets, pipes, open surfaces, flight, etc).

The Reynolds number is the most well-known of a range of dimensionless quantities (see section 4.2), which are at the heart of dimensional analysis [4]. Table 8.1 contains examples of Reynolds numbers; demonstrating the vast range of Re encountered. Within the paradigms of fluid dynamics, various essential quantities, for example, drag forces proportional to velocity, and Poiseuille's law for the flow velocity through tubes and blood vessels, can be derived in the laminar flow regime. But due to the nonlinearity of the fundamental equations, this approach does not give results in the turbulent limit. In such situations, dimensional analysis is especially useful, in that much of the behaviour can be revealed without solving the nonlinear differential equations.

8.2 Drag forces

Consider the dynamics of objects in fluids (in air, or in liquids); in particular, the drag force that arises by the interaction of the object and its supportive medium (where the object moves in a static fluid, or the fluid moves around a static object). Like friction, the drag force always opposes the motion of an object. Unlike simple friction, however, the drag force is proportional to some function of the velocity of the object in the fluid. This functionality is complicated and depends upon the shape

Table 8.1. Range of Reynolds numbers encountered in science and technology.

Moving object	Reynolds number
Ocean liner (length: 100 m), moving at 30 m s^{-1}	3 000 000 000
A whale swimming at 10 m s^{-1}	300 000 000
A tuna swimming at 10 m s^{-1}	30 000 000
A swimmer (2 m), moving at 1 m s^{-1}	2 000 000
A duck flying at 20 m s^{-1}	300 000
A dragonfly moving at 7 m s^{-1}	30 000
A small crustacean moving at 0.2 m s^{-1}	300
Flapping wings of the smallest flying insect	30
An invertebrate larva, 0.3 mm long, moving at 1 mm s^{-1}	0.3
A sea urchin sperm advancing the species at 0.2 mm s^{-1}	0.03
A bacterium swimming at 0.01 mm s^{-1}	0.000 01

and size of the object, its velocity and the fluid medium. For most large objects such as airplanes, cyclists, cars and balls not moving too slowly, the magnitude of the drag force, F_d, is proportional to the square of the speed of the object, (v); that is, $F_d \propto v^2$. When taking into account other factors, this relationship becomes:

$$F_d = \tfrac{1}{2} C_d \rho A v^2, \tag{8.4}$$

where C_d is the drag coefficient, which is a constant of proportionality, A is the area of the object facing the fluid, and ρ is the density of the fluid. The drag coefficient defines how the shape impacts on the force that could be exerted.

8.2.1 Similitude

The size of the object falling through air presents an interesting application of air drag. If you fall from a 5-m-high branch of a tree, you will likely fracture a bone. However, a small squirrel or a cat can do this without injury. You do not reach a terminal velocity in such a short distance, but the squirrel does. The following quote on animal size and terminal velocity is from a 1928 essay (*On Being the Right Size*) by the British biologist, J B S Haldane [5]. '*To the mouse and any smaller animal,* [gravity] *presents practically no dangers. You can drop a mouse down a thousand-yard mine shaft; and, on arriving at the bottom, it gets a slight shock and walks away, provided that the ground is fairly soft. A rat is killed, a man is broken, and a horse splashes. For the resistance presented to movement by the air is proportional to the surface of the moving object. Divide an animal's length, breadth, and height each by ten; its weight is reduced to a thousandth, but its surface only to a hundredth. So, the resistance to falling in the case of the small animal is relatively ten times greater than the driving force.*'

Athletes as well as car-designers seek to reduce the drag force to lower their race times. Aerodynamic shaping of an automobile can reduce the drag force, and thus increase a car's economic efficiency. For a particular design, the value of the drag coefficient, C_d, is determined empirically, usually in a wind tunnel. The drag coefficient can depend upon velocity, but we usually assume that it is a constant. Table 8.2 lists some typical drag coefficients for a variety of objects. At high-speeds, over 50% of the power of a car is used only to overcome air drag on a highway. The most fuel-efficient cruising speed is just below 50 mph (75 kph)—see section 8.2.4.

A great deal of research goes into minimizing drag in the world of competitive sport. The dimples on golf-balls are being redesigned (see section 7.3), as are the costumes that athletes wear. Bicycle racers and some swimmers and runners wear full bodysuits. Such innovations can cut away milliseconds in a race; sometimes making the difference between a gold and a silver medal.

8.2.2 The calculus of velocity-dependent frictional forces

When a body slides across a surface, the frictional force is approximately constant. This is termed Coulomb friction, named after Charles-Augustin de Coulomb; it is an approximate model used to calculate the force of dry friction. It is governed by the model $F_f \leqslant \mu F_n$, where F_f is the force of friction exerted by each surface on the other.

Table 8.2. Some drag coefficients.

Object	Drag coefficient (C_d)
Airfoil	0.05
Sphere	0.45
Bicycle	0.90
Skydiver (falling horizontally)	1.0
Generic family car	0.3–0.4

This frictional force is parallel to the surface, in a direction opposite to the net applied force, μ is the coefficient of friction, which is an empirical property of the contacting materials, and F_n is the normal force exerted by each surface on the other, directed perpendicular (normal) to the surface. The frictional force on a body moving through a liquid or a gas does not behave so simply. This drag force is generally a complicated function of the body's velocity. However, for a body moving in a straight line at moderate speeds (v) through a liquid such as water, see figure 8.2, the frictional force, F_d, can often be approximated by $F_d = -bv$, where b is a constant whose value depends on the dimensions and shape of the body (see equation (8.4)) and the properties of the liquid, and v is the velocity of the body (v has dimensions LT^{-1}, so b must have dimensions of MT^{-1} to generate a force, MLT^{-2}). Let us consider a body of mass, m, falling through a resistive fluid (see figure 8.2) with a velocity v.

Newton's second law in the vertical direction gives the differential equation,

$$mg - bv = m\frac{dv}{dt}, \tag{8.5}$$

where the acceleration is $\frac{dv}{dt}$. As v increases, the frictional force or drag force, $-bv$ increases until it matches mg. At this point, there is no acceleration and the velocity remains constant at the terminal velocity, v_t. From equation (8.5), $mg - bv_t = 0$, so $v_t = mgb$. We can find the object's velocity by integrating the differential equation for v. First, we rearrange terms to obtain

$$dv/g - (b/m)v = dt. \tag{8.6}$$

Assuming that $v = 0$ at $t = 0$, integration of this equation yields

$$\int_0^v dv'/g - (b/m)v' = \int_0^t dt', \text{ or } -\frac{m}{b}\ln(g - \frac{b}{m}v')|^{v_0} = t'|^{t_0},$$

where v' and t' are dummy variables of integration. With the limits given, we find

$$-\frac{m}{b}[\ln(g - \frac{b}{m}v) - \ln g] = t. \tag{8.7}$$

Since $\ln A - \ln B = \ln(A/B)$, and $\ln(A/B) = x$ implies $e^x = A/B$, we obtain:

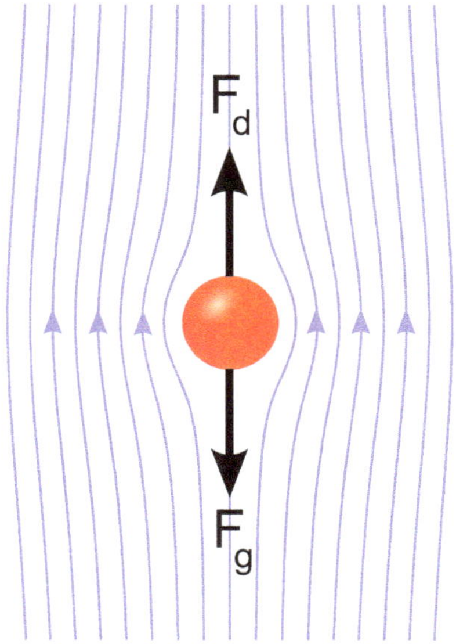

Figure 8.2. Flow past a falling (at a velocity v) sphere (of radius r, and density ρ) in a fluid (of density σ, and viscosity μ). This could be a skydiver falling under the influence of gravity (F_g), with the drag force (F_d) countering his/her downward acceleration. In molecular terms, however, this could be a representation of Stokes' law; where the force countering the gravitational sedimentation ($\propto r^3\rho g$) would include a buoyancy term ($\propto r^3\sigma g$) and the drag force ($\propto rv\mu$), which arises from the viscosity of the supporting fluid. This 'Stokes sphere' image has been obtained by the author from the Wikimedia website where it was made available under a CC BY-SA 3.0 license. It is included within this article on that basis. It is attributed to Kraaiennest.

$$\frac{g - \left(\frac{bv}{m}\right)}{g} = \mathrm{e}^{-\mathrm{bt/m}}, \text{ and } v = \frac{mg}{b}(1 - \mathrm{e}^{-\mathrm{bt/m}}). \tag{8.8}$$

Notice that as $t \to \infty$, $v \to mg/b = v_t$, which is the terminal velocity. The position at any time may be found by integrating the equation for v. With $v = \frac{dy}{dt}$,

$$dy = \frac{mg}{b}((1 - e^{-bt/m})dt. \tag{8.9}$$

Assuming $y = 0$ when $t = 0$,

$$\int_0^y dy' = \frac{mg}{b}\int_0^t \left(1 - exp-\frac{bt'}{m}\right)dt', \tag{8.10}$$

which integrates to

$$y = \frac{mg}{b}t + (m^2g/b^2)e^{-bt/m} - 1. \tag{8.11}$$

The skydiver's mass = 90 kg; that is, weight = 882 N (mg)			
$F_d = 0$	$F_d = 300$ N	$F_d = 600$ N	$F_d = 882$ N
-	↑	↑	↑

↓	↓	↓	↓
$F_g = -882$ N	$F_g = -882$ N	$F_g = -882$ N	$F_g = -882$ N
$a = -9.8$ m/s^2	$a = -6.5$ m/s^2	$a = -3.1$ m/s^2	$a = 0$ m/s^2

Figure 8.3. Dynamics of a person with a mass of 90 kg during a skydive. The initial speed is zero, so the drag force is also zero. As his/her speed increases under the influence of gravity, the drag force (F_d) also grows, eventually cancelling out the person's weight (F_g). At that point, acceleration is zero and terminal velocity is achieved. This 'A parachutist above Venezuela' image has been obtained by the author from the Wikimedia website where it was made available under a CC BY-SA 3.0 license. It is included within this article on that basis. It is attributed to Radikaltech.

To investigate the dynamics of a falling object, the alternative to the above exercise in calculus is to apply dimensional analysis. Whether we are considering an object falling through a viscous fluid, or a skydiver falling from an airplane; in both cases we are interested in the terminal velocity: the velocity that the object reaches after falling for a sufficiently long period, when a balance of dynamical forces is achieved—see figure 8.3.

8.2.3 Dimensional analysis of the drag force

To use dimensional analysis to find the drag force, F_d, the first step is to choose the relevant variables. We begin by including the variable of interest, F_d. What characteristics of the object of interest (we assume a sphere to simplify geometric considerations) are relevant to the drag force? The drag force is independent of the content of the sphere, of radius R. It is primarily a geometrical problem, and the density of the sphere is, to first approximation, negligible. What characteristics of the fluid are relevant? Certainly, its density, ρ.

As noted above, viscosity is a measure of the tendency of a fluid to average out velocity differences in the flow. You can observe an analogue of viscosity in traffic flow on a multilane motorway. If one lane moves much faster than the others, drivers switch from the slower to the faster lane; eventually, slowing down the faster lane. The decisions of the individual drivers reduce the overall velocity gradient. In a fluid, molecular motion transports speed (that is, momentum) from fast- to slow-flowing regions of the fluid. This transport reduces the velocity difference between the regions. Thicker, more viscous fluids cause more drag than thin fluids. So, we include viscosity in our list of variables.

Fluid mechanicians have defined two viscosities, dynamic viscosity μ, and kinematic viscosity υ. They are related by the density (ρ) of the fluid: $\mu = \rho_{\text{fluid}}\upsilon$. Vogel [6] discusses these two types of viscosity in detail, and their huge importance for small animals and insects. For the analysis of drag force, we need only know that viscous forces are proportional to viscosity. But which viscosity should we use? Dynamic viscosity hides ρ_{fluid} inside the product $\upsilon\rho_{\text{fluid}}$; equations that contain ρ_{sphere} and μ are difficult to form into dimensionless groups—as there are no other variables with mass (see table 8.3 for all the variables in this problem) with which they may be combined. So, we use the kinematic viscosity, υ. Thus, we may write our initial hypothesis about the nature of the drag force:

$$F_{\text{d}} \propto (R)^a(\upsilon)^b(v)^c(\rho_{\text{fluid}})^d = K(R)^a(\upsilon)^b(v)^c(\rho_{\text{fluid}})^d, \tag{8.12}$$

where K is a constant of proportionality.

The drag force does not depend on ρ_{sphere} or g (acceleration due to gravity); it is simpler to estimate than the terminal velocity (which does depend on g and ρ_{sphere}). The five variables (we are not using the dynamic viscosity) in the list are composed of three base dimensions (M, L and T); so, by the Buckingham Π-theorem we are looking for (5–3) two dimensionless groups of variables. The list already includes a velocity, v. If we can construct another quantity, V, also with dimensions of velocity, then we can form the dimensionless group v/V. The viscosity υ almost works; it has an extra power of length, which R can eliminate. So, we have: $V \equiv \upsilon/R$. The first dimensionless group is then

$$\Pi_1 \equiv \frac{v}{V} = \frac{vR}{\upsilon} = \frac{\rho v R}{\mu}, \tag{8.13}$$

which is the Reynolds number, Re. It is a dimensionless measure of the flow speed. The velocity alone cannot distinguish fast from slow flow, because v is not dimensionless. If you hear that a quantity is small, or fast, or large, or whatever, your first reaction should be 'compared to what?' To distinguish a fast flow from a slow flow, we have to use a dimensionless quantity related to v; the Reynolds number. It compares v to V. Low values of Re indicate slow, viscous flow (cold honey); high values indicate turbulent flow (a jet flying at 600 mph)—see table 8.1.

Table 8.3. Variables for calculating drag force.

Variable	Dimension	Description
F_{d}	MLT^{-2}	Drag force
R	L	Radius of falling sphere
υ	L^2T^{-1}	Kinematic viscosity
ρ_{fluid}	ML^{-3}	Fluid density
v	LT^{-1}	Terminal velocity
μ	$\text{ML}^{-1}\text{T}^{-1}$	Dynamic viscosity

For the second dimensionless group, as the drag force is absent from Π_1, it has to be part of Π_2. Instead of attempting to derive the dimensionless group by inspection, we construct two quantities with the same units, and divide them to obtain a dimensionless quantity. Notice that F_d R has dimensions of energy ($\mathsf{MLT^{-2}}.\ \mathsf{L} = \mathsf{ML^2T^{-2}}$). We construct another energy (call it E), then form the dimensionless ratio $F_d R/E$. Energy contains one power of mass; the only variable other than F_d that contains mass is ρ_{fluid}. So, E contains one power of ρ_{fluid}. We start with $\rho_{\text{fluid}}R^3$, which has dimensions of mass ($\mathsf{ML^{-3}\ L^3}$). Kinetic energy is $\frac{1}{2}mv^2$, so it has units of $\mathsf{M}[V]^2 = \mathsf{M}\ (\mathsf{LT^{-1}})^2$. The second energy is therefore:

$$E \equiv \rho_{\text{fluid}} R^3 v^2 (\text{dimensionally, we have } \mathsf{ML^{-3}L^3(LT^{-1})^2} = \mathsf{ML^2T^{-2}}). \tag{8.14}$$

The second dimensionless group, Π_2, is the ratio of the two energies:

$$\Pi_2 \equiv F_d R/E = F_d/\rho_{\text{fluid}} R^2 v^2. \tag{8.15}$$

From the Buckingham Π-theorem, $f(\Pi_1,\Pi_2) = 0$, we want to solve for F_d, which is contained in Π_2, so we use $\Pi_2 = f(\Pi_1)$. In terms of the original variables, the drag force is

$$F_D = \rho_{\text{fluid}} R^2 v^2 f\left(\frac{\rho v R}{\mu}\right). \tag{8.16}$$

To learn the form of the function f, we look at our problem at limiting conditions: at turbulent, high-speed flow (Re $\gg 1$), or at viscous, low-speed flow (Re $\ll 1$)—see table 8.1.

8.2.4 A car on a highway

The drag equation gives the force experienced by an object moving through a fluid at relatively large velocity (high Reynolds number, Re > ~1000). This is also called quadratic drag (it is quadratic in the velocity relevant to the situation). The drag equation, equation (8.4), is attributed to Lord Rayleigh, who originally used L^2 in place of the objects' area, A (L being a length variable). It is readily shown that for a car on a highway, travelling with a velocity, v, in air of density, ρ, and orthographic projection (frontal area), A, that the drag force, F_d, is given by: $F_d = f(\rho, v, A) = [\rho]^x [v]^y [A]^z$; that is,

$$\mathsf{MLT^{-2}} = (\mathsf{ML^{-3}})^x(\mathsf{LT^{-1}})^y(\mathsf{L^2})^z \rightarrow \mathsf{M}^x\mathsf{L}^{-3x+y+2z}\mathsf{T}^{-y};$$

giving: $F_d = C\rho v^2 A$, where C is a constant of proportionality. This drag depends on the properties of the fluid, and on the size, shape and speed of the object. Above, we encountered the drag coefficient, C_d, a dimensionless property of the object under investigation (see table 8.2). The drag coefficient depends on the shape of the object and on the Reynolds number, Re.

Under the assumption that the fluid into which the car is travelling is not moving relative to the reference system, the car, the power required to overcome the aerodynamic drag (P_d) at higher speeds is given by:

$$P_d = F_d v = \tfrac{1}{2}\rho v^3 A C_d.$$

Dimensionally, this would be: $ML^2T^{-3} = MLT^{-2}\,LT^{-1} = ML^{-3}\,L^3T^{-3}\,L^2$ (C_d being without dimension). Note that the power needed to push an object through a fluid increases as the cube of the velocity. Consequently, a car cruising on a highway at 50 mph may require only 10 horsepower (7.5 kW) to overcome aerodynamic drag, but that same car at 100 mph requires 80 horsepower (60 kW). With a doubling of the speed, the drag force experienced quadruples as per the equation. Exerting four times the force over a fixed distance produces four times as much work. Yet at twice the speed, the work (resulting in displacement over a fixed distance) is done twice as fast. Since power is the rate of doing work, four times the work done in half the time requires eight times the power.

When the fluid is moving relative to the reference system (a car driving into a strong headwind) the power required to overcome the aerodynamic drag is given by:

$$P_d = F_d v_0 = \tfrac{1}{2}\rho A C_d (v_{wind} + v_0)^2 v_0,$$

where v_{wind} is the wind speed and v_0 is the object speed (both relative to ground).

8.2.5 The skydiver

Computing the terminal velocity of a skydiver allows us to make further observations about drag. Consider a skydiver falling under the influence of gravity. The two forces acting on him/her are the force of gravity, and the drag force (ignoring the small buoyant force, which is important if we are considering molecules or snowflakes, but not skydivers). The downward force of gravity remains constant regardless of the velocity of the skydiver. However, as the skydiver's velocity increases, the magnitude of the drag force increases until the magnitude of the drag force is equal to the gravitational force, thus producing a net force of zero; this is illustrated in figure 8.3. A zero, net force means that there is no acceleration—as defined by Newton's second law; that is, $F = ma = 0$. At this point, the skydiver's velocity remains constant, and we say that the skydiver has reached terminal velocity, v_t. At terminal velocity, the net force is given by: $F_{net} = F_d—mg = ma = 0$.

Using the equation for drag force (equation (8.4)), we have

$$mg = \tfrac{1}{2}C_d\,\rho A v_t^2. \tag{8.17}$$

Solving for the velocity, we obtain

$$v_t = \sqrt{2mg\rho C_d A}. \tag{8.18}$$

Depending upon the orientation of the skydiver; that is, the factor A, the terminal velocity can vary from 350 km per hour (69.4 ms^{-1}) while falling in a pike (head first) position, to about 200 km per hour (55.5 ms^{-1}) in a spread-eagle position (see figure 8.3).

While the transition from laminar to turbulent flow occurs at a Reynolds number of approximately 2300 in a pipe, the precise value for the transition depends on small perturbations. If the experiment is carefully arranged so that the pipe is smooth and there are no disturbances to the velocity, higher values of Re can be obtained with the flow still in a laminar state. However, if Re is less than 2300, the flow will be laminar even if it is disturbed. Thus, 2300 is the value of Re below which turbulence will not occur in a pipe. Moreover, if the flow has a different geometry, such as flow in a square duct, or over a turbine blade, transition will occur at different values of Re. The essential point is that flows become turbulent at high Reynolds numbers where 'high' means much greater than unity. The Reynolds number of a falling skydiver is large. If we assume that he/she tucks into a foetal position, becoming roughly spherical, with, say, $R \sim 0.5$ m, as a person is mostly water, so $\rho_{\text{skydiver}} \sim 1 \text{ g cm}^{-3}$. The density of air is $\rho_{\text{air}} \sim 10^{-3} \text{ g cm}^{-3}$, so $\rho_{\text{air}} \ll \rho_{\text{skydiver}}$; buoyancy is not an important effect. Using these conditions, we calculate a terminal velocity of about 60 ms^{-1}; however, this calculation assumed Re $\gg$ 1. The actual value will be, of order, 10^5—see table 8.1.

8.2.6 Stokes' law

If, however, the velocity of the body is sufficiently small, we expect a steady, laminar flow around the body. In the reference frame of the body, the velocity at every location is constant and therefore the macroscopic kinetic energy is constant. This means that the work done by the pulling force must go entirely into thermal energy, caused by the internal friction in the fluid. In this limit, the drag force will therefore depend on the viscosity of the fluid but not on its density, because the density enters into only the macroscopic kinetic energy.

Thus, in the low-velocity limit, we may write (variables are listed in table 8.3):

$$F_{\text{d}} \propto (R)^a (v)^b (\mu)^c = K'(R)^a (v)^b (\mu)^c,$$

where K' is a constant of proportionality. We are now using the dynamic viscosity, μ. Taking the dimension of both sides gives

$$\mathsf{MLT}^{-2} = \mathsf{M}^c \mathsf{L}^{a+b-c} \mathsf{T}^{-b-c}.$$

The unique set of solutions are: $a = b = c = 1$. The formula for the drag force in this limit of low velocity is

$$F_{\text{d}} = K'\mu R v, \tag{8.19}$$

where the constant will depend upon of the shape of the body. For a sphere of radius, R, the number is 6π, and equation (8.19) is known as Stokes' law, after the Anglo-Irish physicist, Sir George Gabriel Stokes (1819–1903).

In the opposite limit, where the velocity is sufficiently large, we expect a turbulent wake behind the body. When pulling the body through the fluid, we continuously produce new macroscopic kinetic energy in the wake. At sufficiently high velocities, we assume that the work done by the pulling force is converted predominantly into macroscopic kinetic energy in the wake. The macroscopic kinetic energy of a fluid, with a given flow pattern depends on the density of the fluid, but not on its viscosity. Although the viscosity of the fluid determines how the macroscopic kinetic energy in the wake is transformed into thermal energy in the long run, the immediate production of thermal energy in the fluid can be neglected. In this case, we therefore expect the pulling force (and the oppositely directed drag force) to be independent of viscosity; determined only by shape, v, R and ρ. Again, we now have only three input variables, besides the dimensionless shape variables, so the dependence of F_d on the three variables is determined uniquely via the Rayleigh algorithm. By an argument entirely analogous to those given previously, we find

$$F_d = K\,\rho R^2 v^2. \tag{8.20}$$

This formula describes the air resistance acting on a bicyclist, a falling skydiver and a moving motor vehicle. Since μ/ρ is 15×10^{-6} m^2 s^{-1} for air at room temperature, typical values such as $R = 1.5$ m and $v = 10$ ms^{-1} give a Reynolds number of order 10^6.

Comparing equation (8.19) valid for small velocities, with equation (8.20), which is valid in general, we conclude that $f(\mathrm{Re})$ is proportional to $(\mathrm{Re})^{-1}$ at small velocities. Similarly, by comparing equation (8.20), valid for large velocities, with equation (8.15), we find $f(\mathrm{Re})$ = constant. Since $f(\mathrm{Re})$ is a function of Re, and not of v, we should rather state that $f(\mathrm{Re}) \propto (\mathrm{Re})^{-1}$ at small values of Re, and that $f(\mathrm{Re})$ is independent of Re at large values of Re. In the reasoning above, r, g and ρ were assumed constant and only v varied. But the more general statements are that equation (8.19) is valid in the limit of small Reynolds numbers, and equation (8.20) is valid in the limit of large Reynolds numbers. How small or large the Reynolds numbers should be in order to make equations (8.19) and (8.20) reliable depends on the shape of the body. In the cases of bicycling, car driving and falling heavy objects, equation (8.20) is valid and F_d is thus independent of Re. This means that the assumption of F_d being independent of g is justified in these cases. In general, at a given Re, equation (8.20) is more valid for not-very-streamlined bodies, where the cross-section of the wake is determined by the cross-section of the body, not by the detailed streaming of the fluid around the body.

Good examples of Stokes' law are provided by microorganisms, pollen and dust particles (see table 8.1); and it was used by Einstein in his analysis of Brownian motion [7]. Because each of these objects is so small, we find that many of these objects travel unaided only at a constant (terminal) velocity. Terminal velocities for bacteria (size about 1 μm) can be about 2 μm s^{-1}. To move at a greater speed, many bacteria swim using flagella (organelles shaped like little tails). If we compare animals living on land with those in water, you can see how drag has influenced evolution [6]. Fish, dolphins and even massive whales are streamlined in shape to reduce drag forces. Birds are streamlined and migratory species that fly large distances often have particular features such as long necks. Flocks of migrating birds fly in the shape of a spearhead as the flock forms a streamlined pattern.

8.3 Bubbles in fizzy drinks

Let us consider another example of the use of dimensional analysis to describe a physical system, which at first sight may seem complex, and perhaps intractable; but which also tells us something about the dynamics of objects in fluids. I am sure that we have all, at some time stared at a glass of a fizzy beverage watching the bubbles rise in the fluid. In addition, I am sure that we have noticed that some bubbles rise faster than others. So, let us consider the velocity of the bubbles as a function of their size. We wish to find the velocity, v, as a function of the bubble diameter, d[1].

We know that $[v] = \mathsf{LT}^{-1}$, and we seek independent variables with dimensions involving L and T, so as to construct dimensionless groups. These independent quantities are bubble size (diameter of bubble), $[d] = \mathsf{L}$, and the acceleration due to gravity, $[g] = \mathsf{LT}^{-2}$. The quantity of interest is the force on the rising bubbles, f_b; that is $[f_b] = \mathsf{MLT}^{-2}$. The variables for this problem are given in table 8.4.

We can say immediately that v = constant $x\sqrt{(gd)}$. This can be seen from the balance of two forces (and represents another dimensionless quantity, the Froude number, Fr; see section 4.1). The force of buoyancy (F_B) due to Archimedes is: $F_B \sim \rho d^3 g$; $[F_B] = \mathsf{ML}^{-3}\,\mathsf{L}^3\,\mathsf{LT}^{-2} = \mathsf{MLT}^{-2}$, which we set equal to the drag force, F_d, $F_d \sim \rho v^2 d^2$; $[F_d] = \mathsf{ML}^{-3}\,\mathsf{L}^2\mathsf{T}^{-2}\,\mathsf{L}^2 = \mathsf{MLT}^{-2}$. Balancing these forces allows us to say $v^2 \sim gd$. However, when we calculated the drag force, we assumed the bubbles are moving at high-speed, a result which neglects the viscosity of the fluid. This will be an unrealistic assumption; especially, for bubbles which rise rather slowly, and at very different speeds in fluids of varying viscosity. Thus, we must also include the effect of viscosity.

Even with these terms we are still making a number of assumptions; for example, the weight of the air in the bubble, the pressure changes with height in the fluid and associated expansion, and the effect of surface tension on the shape of the bubble; these terms are of huge importance in an oil well—if you wish to avoid a blow-out. However, without these small effects (for a glass of champagne) we have from the Buckingham Π-theorem 6 variables (ignoring g), 3 independent base dimensions,

Table 8.4. Variables for the problem of rising bubbles.

Variable	Dimension	Description
ρ	ML^{-3}	Fluid density
f_b	MLT^{-2}	Force on bubble
P	$\mathsf{ML}^{-1}\,\mathsf{T}^{-2}$	Pressure in bubble
v	LT^{-1}	Velocity of bubble
d	L	Diameter of bubble
μ	$\mathsf{ML}^{-1}\mathsf{T}^{-1}$	Viscosity of fluid
g	LT^{-2}	Acceleration due to gravity

[1] This is a general case of a problem the author was asked to solve in his time as a mathematical modeller with an oil exploration company—but instead of glasses of fizzy beverages, the bubbles were forming at the bottom of an oil well, and we wished to know something about their dynamics as they moved up the well-bore; from a place of high-pressure and temperature to the surface at ambient temperature and pressure. As a rule of thumb, in a well-bore one assumes that the pressure increases by ½ psi/foot of depth—try your skill at putting that into an SI unit using the method outlined in chapter 3.

and so 3 dimensionless combinations of variables. The first dimensionless parameter, Π_1, is related to the Mach number (Ma) = $v/\sqrt{(P/\rho)}$.

For the second dimensionless combination, Π_2, consider,

$$[\mu] = \mathrm{ML^{-1}T^{-1}} = [\rho]^a[v]^b[d]^c = \mathrm{M}^a\mathrm{L}^{-3a+b+c}\mathrm{T}^{-b}, \tag{8.21}$$

and with $a = b = c = 1$, we define the dimensionless Reynolds number (Re), $\mathrm{Re} = \frac{\rho v d}{\mu}$. The final nondimensional group, Π_3, is the force on a rising bubble, f_b, which is $\rho v^2 d^2$. Now, we apply the Buckingham Π-theorem, $F(\Pi_1, \Pi_2, \Pi_3) = 0$, to see that the force on the rising bubbles, f_b, is:

$$f_\mathrm{b}/\rho v^2 d^2 = F(\mathrm{Ma},\ \mathrm{Re}). \tag{8.22}$$

Here, the expression is in terms of the Reynolds number and Mach number. This expression defines what it means to move fast enough to ignore friction. To do so we need Re >> 1 and assume that $F(\mathrm{Ma}, \infty)$ is finite. For real bubbles, however, we first note that bubbles move much more slowly than the speed of sound, so can set Ma = 0, leaving us only the dependence on Re to determine. The familiar limit is where Re >> 1, and so the force becomes:

$$f_\mathrm{b} = \mathrm{constant}.\ \rho v^2 d^2, \tag{8.23}$$

where this constant is $F(0, \infty)$. Physically, this situation is the case where the fluid behind the bubble becomes turbulent and viscous forces no longer matter. This is termed turbulent drag.

In the opposite limit, Re << 1, it is clear that F(0, 0) is not finite, or our answer would be independent of viscosity, and we know that if viscosity were infinite the viscous force would be so large that our bubbles would be immobile (the bubbles would be immobilised in a gel). So, to consider what $F(\mathrm{Re})$ looks like at small Re, we need some physical insight, and/or some experimental data.

If the viscosity is large, we could argue the drag force is independent of the density, as the inertia of the fluid is no longer important. Therefore, since the force is $f_\mathrm{b} = \rho v^2 d^2\ F(\mathrm{Re})$, the way to cancel the dependence on the density ρ is by setting

$$\mathrm{F(Re)} = \frac{\mathrm{constant}}{\mathrm{Re}}, \tag{8.24}$$

which leaves us with the limiting case of Stokes' formula for drag force, f_b = constant × $\mu v d$, which applies for slow velocities, large viscosities, or low densities/small bubbles. Thus, we can once again apply the force balance equation with the Archimedes force:

$$\rho d^3 g \sim \mu v d \rightarrow v = \mathrm{constant}\frac{\rho g}{\mu}d^2. \tag{8.25}$$

Therefore, the dependence of bubble velocity on the diameter is as seen. Note that we go through an intermediate region which can depend on all sorts of complications to do with bubble shape changes, surface tension and pressure changes.

The crossover occurs where Re ~ 1, and we are unable to make either of our previous approximations.[2] These different regimes of drag forces (as represented by the drag coefficient) and Reynold's numbers are shown in figure 8.4.

The fluid flow around spherical pollen particles, microbes, or bubbles is strongly dependent on the particle's Reynolds number. The flow at Re « 1 is called the Stokes' regime (or creeping flow—position No.2 on the curves in figure 8.4); where the flow inertial terms are negligible with respect to viscous terms and flow remains attached to the sphere with no trailing wake. The flow remains attached up to Re ≈ 20, which is the onset of flow separation. At 20 < Re < 130 circular wakes behind a sphere grow but they remain steady and attached to the particle (position No.3 on the curves in figure 8.4). As Re increases beyond 130 and up to 1000, vortex shedding begins and wakes behind the sphere gradually become unstable and unsteady. At $1000 < \text{Re} < 3 \times 10^5$ wakes behind the sphere become fully turbulent while the boundary layer at the front of the sphere is laminar. This range of Reynolds number is called the Newton regime. $\text{Re} > 3 \times 10^5$ corresponds to critical transition and supercritical regime where boundary layer and wake behind the sphere are both turbulent. $\text{Re} = 3 \times 10^5$ is called the critical Reynolds number, at which the drag crisis occurs and reduces the drag coefficient markedly (position No.5 on the curves in figure 8.4).

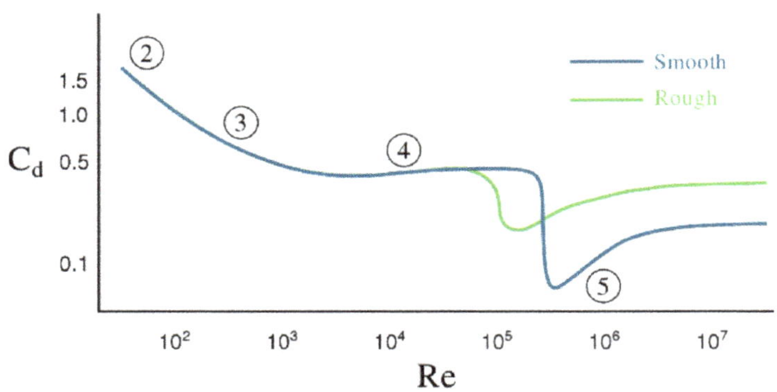

Figure 8.4. Drag coefficient C_d for a sphere as a function of Reynolds number, Re, as obtained from laboratory experiments. The blue line is for a sphere with a smooth surface, while the green line is for the case of a rough surface. The numbers along the line indicate several flow regimes and associated changes in the drag coefficient: ②: attached flow (Stokes flow) and steady separated flow, ③: separated unsteady flow, having a laminar flow boundary layer upstream of the separation, and producing a vortex street, ④: separated unsteady flow with a laminar boundary layer at the upstream side, before flow separation, with downstream of the sphere a chaotic turbulent wake, ⑤: post-critical separated flow, with a turbulent boundary layer. The effect occurs at lower Reynolds numbers when the ball/bubble is rough (such as a golf-ball with dimples, see section 7.3) than when it is smooth. (Image reproduced from NASA. Image stated to be in the public domain. It is included within this article on that basis.)

[2] It is generally the case that when a dimensionless number approaches 1, interesting physics becomes apparent.

8.4 Magnetic-braking: the terminal velocity of a magnet falling in a tube of a non-magnetic metal

As a final example of a drag force, we will look at the magnetic-braking of a cylindrical, permanent magnet falling inside a metal pipe of a non-magnetic metal—usually of aluminium or copper [9–12]. This is an experiment frequently shown in lecture demonstrations (see YouTube video at [13]). It is an experiment that captures the students' and public's imagination; however, it is a phenomenon that is not straightforward to explain. We will examine some of the details of the slowing down of permanent magnets falling in tubes made of non-magnetic metals, and use dimensional analysis to attempt to unravel something of the underlying physics. However, as with falling skydivers and bubbles rising in a glass of champagne, the terminal velocity of a magnet falling in a conducting tube is about a drag force being generated between the permanent magnet and the electron density of the metal tube. A drag force can become sufficiently important to counter-balance the mass of the permanent magnet (of a dense alloy) falling in the Earth's gravity field.

Consider a copper pipe of internal radius a, and wall thickness w. The experimental demonstration is straightforward: a cylindrical, permanent magnet of cross-sectional radius slightly smaller than a, falls down the bore of the tube and is seen to slow (see the video in [13]). It is necessary to imagine that the pipe is uniformly subdivided into parallel rings of width, l. When the magnet is released, the magnetic flux in each one of the rings begins to change. In accordance with Faraday's law, the magnetic flux induces an electromotive force, and an electric current inside the ring of metal. The magnitude of the current will depend on the distance of each ring from the falling magnet, as well as the velocity of the magnet. The Biot–Savart law states that an electric current produces its own magnetic field which, according to Lenz's law, must oppose the action that induced it; that is, the motion of the magnet. Thus, if the magnet is moving away from a given ring, the induced field will try to attract it back; and the magnet will simultaneously be moving towards the next ring where the induced field will tend to repel it. The net force on the magnet can be calculated by summing the magnetic interaction of the magnet with all the rings. The electromagnetic force is an increasing function of the velocity, and will decelerate the falling magnet. When the fall velocity reaches the value at which the magnetic force completely compensates gravity, acceleration will go to zero, and the magnet will continue falling at a constant terminal velocity v_t. The terminal velocity is achieved more rapidly for a strong magnet, than for a weaker magnet of the same mass.

This interaction between the magnet and the metal tube is a manifestation of magnetic induction, described by Faraday's law of induction. We need expressions for both the magnitude of the magnetic drag force, F_d, on the magnet, and the induced EMF, ε_i, on the infinitesimal conductive rings into which the tube length has been divided. For a small ring of metal of radius, a, moving with a small velocity, v, in the field, $\boldsymbol{B}$, of a magnet, the induced EMF is given by

$$\varepsilon_i = \int (v \times \boldsymbol{B}) \cdot \mathrm{d}\boldsymbol{l} = vB_r(2\pi a), \tag{8.26}$$

where B_r is the radial component of the magnetic field, and the integral is evaluated along the ring. The axial force component F_z opposing the motion of the ring along the z-axis is

$$F_z = I(\boldsymbol{l} \times \boldsymbol{B})_z = 2\pi IaB_r. \tag{8.27}$$

For a magnet falling inside a tube, the field $\boldsymbol{B}$ is produced by the magnet itself, and we consider a small motionless element of the pipe in the shape of a ring of length dz of internal and external radii a and b; that is, thickness w, equal to the pipe radii. This ring carries an induced electric current dI due to the changing magnetic flux. It is thus necessary to obtain an expression for the magnetic field produced by the magnet. This field, $\boldsymbol{B}$, may be approximated as being produced by a simple magnetic dipole (of dimension $\mathsf{L}^2\mathsf{I}$).

Lenz's law arises from Faraday's law of induction, being responsible for the negative sign: $\varepsilon_i = -\partial\Phi B/\partial t$, which indicates that the induced electromotive force, ε_i, and the rate of change in magnetic flux ΦB have opposite signs. Thus, the direction of the back EMF of an induced field opposes the charging current that created it. When an object falls in a gravitational field, its potential energy is converted into kinetic energy. In the case of a falling magnet inside a copper pipe, the situation is different. Since the magnet moves at a constant velocity, its kinetic energy does not change and the gravitational potential energy must be transformed into something else. This is the ohmic heating of the copper pipe. The gravitational energy must, therefore, be dissipated by the eddy currents induced in the pipe. In the steady-state, the rate at which the magnet loses its gravitational energy is equal to the rate of the energy dissipation by the ohmic resistance,

$$mgv = \sum I(z)^2 R. \tag{8.28}$$

Here, z is the coordinate along the pipe length, $I(z)$ is the current induced in the ring located at some position z, and R is the resistance of the ring. Since the time-scales associated with the speed of the falling magnet are much larger than those associated with the decay of eddy currents, almost all the variation in electric current through a given ring results from the changing flux due to the magnet's motion.

Let us first consider the variables involved in estimating the drag force operating on the falling magnet, F_d. The falling magnet is a phenomenon controlled by two forces: gravity and a magnetic-induced drag. We assume the magnetic field-induced drag to be proportional to the terminal velocity, $k\, v_t$; where k is the magnetic damping constant $[k] = \mathsf{MT}^{-1}$ (kg s^{-1}) and v_t is the terminal velocity, LT^{-1}; thus, $k\, v_t$ is a force (MLT^{-2}). The variables involved in this problem are given in table 8.5.

We are seeking the function, $f(k, a, \mu_0, M_0, \sigma, w) = 0$. There are six variables ($n = 6$) listed in table 8.5, composed of four base dimensions ($m = 4$). The number of dimensionless parameters involved in this problem is then given by $n - m = 2$, which would, by the Buckingham Π-theorem, be related to a still unknown function $f(\Pi_1, \Pi_2) = 0$, where Π_1 and Π_2 are the two dimensionless Π-groups to be identified. The original function, with six variables, is replaced by a function with only two nondimensional parameters.

Table 8.5. Variables involved in magnetic-braking.

Variables	Dimension	Description
$k\ v_t$	MLT^{-2}	Drag force (as described in text)
a	L	Radius of the tube interior
μ_0	$MLT^{-2}\ I^{-2}$	Permeability of the vacuum
M_0	L^2I	Magnetic dipole moment (A m^2 = N m/Tesla = J/Tesla)
σ	$M^{-1}L^{-3}T^3\ I^2$	Conductivity of the tube (siemens $metre^{-1}$; $A^2\ s^3\ kg^{-1}\ m^{-3}$)
w	L	Thickness of the tube wall

The two dimensionless groups will be constructed from the two variables left after choosing the repeat variables; namely the magnetic damping constant k, and the thickness of the tube wall w. The four variables used as repeat variables are: μ_0, M_0, a and σ. As the Π-groups are all dimensionless; that is, they have dimensions $M^\circ L^\circ T^\circ I^\circ$, we can use the principle of dimensional homogeneity (the Rayleigh algorithm) to equate the Π-groups.

For the first parameter, $\Pi_1 = (\mu_0)^x\ a^y\ \sigma^z\ (M_0)^t\ k$, we write the dimensional equations as

$$\Pi_1 = (MLT^{-2}I^{-2})^x L^y (M^{-1}L^{-3}T^3I^2)^z (IL^2)^t MT^{-1} = M^\circ L^\circ T^\circ I^\circ. \tag{8.29}$$

For each dimension (M, L, T or I) the exponents must be equal on both sides of the equation, so:

$$M: x - z + 1 = 0$$

$$L: x + y - 3z + 2t = 0$$

$$T: -2x + 3z - 1 = 0$$

$$I: -2x + 2z + t = 0.$$

The solutions are: $x = -2$, $y = 3$, $z = -1$, $t = -2$; giving

$$\Pi_1 = ka^3/\sigma\mu_{0^2}M_{0^2}. \tag{8.30}$$

A similar procedure is followed for the second Π-group; $\Pi_2 = (\mu_0)^x\ a^y\ \sigma^z\ (M_0)^t\ w$, so

$$\Pi_2 = (MLT^{-2}I^{-2})^x L^y (M^{-1}L^{-3}T^3I^2)^z (IL^2)^t L = M^\circ L^\circ T^\circ I^\circ$$

$$= M^{x-z} L^{x+y-3z+2t+1} T^{-2x+3z} I^{-2x+2z+t}. \tag{8.31}$$

The solutions are: $x = 0$, $y = -1$, $z = 0$, and $t = 0$; giving, $\Pi_2 = \dfrac{w}{a}$. (8.32)

The problem of a permanent magnet falling in a non-magnetic metallic tube may be described by the following function of the two dimensionless groups of variables:

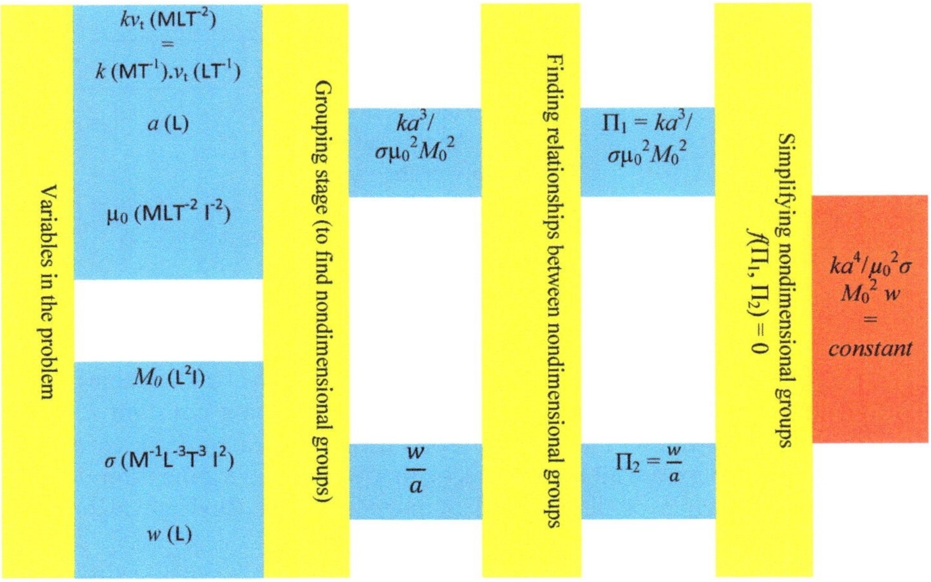

Figure 8.5. Schematic for the dimensional analysis of magnetic-braking. The Π-theorem tells us there are two nondimensional groups that relate the selected variables. These two nondimensional groups of variables are identified and then equated, generating a defining equation.

$$f(\Pi_1, \Pi_2) = 0 \rightarrow f\left(ka^3/\sigma\mu_{0^2}M_{0^2}, \frac{w}{a}\right). \tag{8.33}$$

Once identified, manipulation of the Π-groups is permitted. The magnetic damping constant, k, is proportional to the eddy currents induced in the tube, and these are proportional to the thickness of the tube wall w; then, $f(\Pi_1, \Pi_2) = 0$ may be written as:

$$f(ka^4/\mu_{0^2}\sigma M_{0^2}\, w) = 0, \text{ or } ka^4/\mu_{0^2}\sigma M_{0^2}\, w = C, \tag{8.34}$$

where C is a constant. The manipulations performed are physically plausible since the eddy currents should be proportional to the product σw, as has been theoretically found by Saslow [9]. This dimensional analysis is summarised in figure 8.5.

Given the complexity of fully analysing this problem, the use of dimensional analysis is a more straightforward approach to revealing the relationship amongst the variables that control the phenomenon. Experiments will better define the equations derived from dimensional analysis, and identify the various nondimensional groups that contribute to the phenomenon. What we see above is the bare bones of the phenomenon; the fuller explanations of this effect (see Further reading) shows that there are some rather large numerical factors that dimensional analysis cannot reveal—they are all contained in the factor, C, in equation (8.34).

References

[1] Rayleigh L 1915 The principle of similitude *Nature* **95** 66
[2] Bohr N 1913 On the constitution of atoms and molecules *Philos. Mag.* **26** 1–25
[3] White F M 2016 *Fluid Mechanics* 8th edn (New York: McGraw-Hill)
[4] Bridgman P W 1922 *Dimensional Analysis* (New Haven, CT: : Yale University Press)
[5] Haldane J B S 1985 *On Being the Right Size, and Other Essays* Oxford Paperbacks (Oxford: Oxford University Press)
[6] Vogel S 1994 *Life in Moving Fluids: The Physical Biology of Flow* (Princeton, NJ: Princeton University Press)
[7] Pais A 1982 *Subtle is the Lord: The Science and the Life of Albert Einstein* (Oxford: Oxford University Press) ch 5
[8] Fox R W, McDonald A T and Pritchard P J 2015 *Introduction to Fluid Mechanics* 9th edn (Hoboken, NJ: Wiley)
[9] Saslow W M 1992 Maxwell's theory of eddy currents in thin conducting sheets, and applications to electromagnetic shielding and MAGLEV *Am. J. Phys.* **60** 693
[10] Donoso G, Ladera C L and Martín P 2009 Magnet fall inside a conductive pipe: motion and the role of the pipe wall thickness *Eur. J. Phys.* **30** 855–69
[11] Ireson G and Twidle J 2008 Magnetic braking revisited: activities for the undergraduate laboratory *Eur. J. Phys.* **29** 745–51
[12] Iniguez J, Raposo V, Hernandez-Lopez A, Flores A G and Sazo M 2004 Study of the conductivity of a metallic tube by analysing the damped fall of a magnet *Eur. J. Phys.* **25** 9
[13] https://youtube.com/watch?v=5BeFoz3Ypo4

Chapter 9

Why is the sky blue?

Blue colour is everlastingly appointed by the deity to be a source of delight.
John Ruskin

First, some thoughts about light intensities. Light intensity is defined as: the luminous flux in a certain direction, radiated per unit of solid angle. In the SI, a solid angle is expressed in a dimensionless unit named a steradian (Sr); equal to the angle at the centre of a sphere subtended by a part of the surface equal in area to the square of the radius (see figure 9.1) [1].

In general, a light-source will not radiate its luminous flux uniformly in all directions. If, however, we imagine a sufficiently narrow cone, with its vertex at the light-source, then the luminous flux contained in this cone will approximate a uniform distribution. The light intensity within this narrow cone can be defined as the luminous flux divided by the opening of the cone, expressed in terms of the solid angle of the cone. The result is called the luminous intensity (I), measured in candela (cd), in the direction of the centre-line of the cone. It is not possible to measure the luminous intensity directly. In order to derive the value of the luminous intensity you have to measure the illuminance (the total luminous flux incident on a surface, per unit area; symbol, E_v, and with dimensions L^{-2} J; that is, cd Sr m^{-2}) and calculate it via the inverse square law.

9.1 Quantifying light intensity: subjectively and in absolute terms

The candela is defined in the SI to be the luminous intensity in a given direction, of a source that emits monochromatic radiation of frequency 540×10^{12} hertz (in the green), and that has a radiant intensity in that direction of 1/683 watt per steradian. The luminous efficacy of monochromatic radiation of frequency 540×10^{12} Hz, K_{cd}, is 683 lm/W [1]. (See table 1.3 for the definitions of all the base units of the SI.)

doi:10.1088/978-0-7503-3655-0ch9

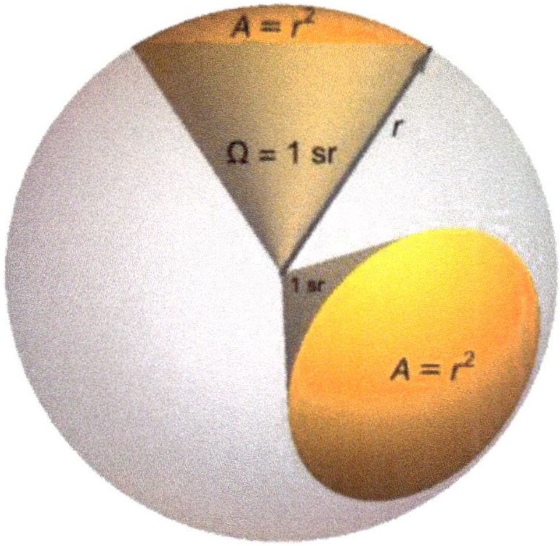

Figure 9.1. A solid angle relates a portion of the volume of a sphere to the surface area it subtends. If that area equals the sphere's radius squared, the solid angle is one steradian. This diagram displays two solid angles of one steradian, viewed from different directions. This 'Solid Angle' image has been obtained by the author from the Wikimedia website where it was made available under a CC BY-SA 4.0 license. It is included within this article on that basis. It is attributed to Andy Anderson.

The lumen (lm) is the unit of luminous flux. It is defined in terms of candela steradians (cd Sr). One lumen is the amount of light emitted in a solid angle of 1 Sr, from a source that radiates to an equal extent in all directions, and whose intensity is 1 cd. One lumen is the equivalent of 1.46 milliwatt of radiant electromagnetic power at a frequency of 540×10^{12} Hz. In SI base units, one lumen is equal to 1.46×10^{-3} kg m^2 s^{-3}. The definitions of light intensity can vary widely—there is no direct measurement, and so the description is experiment dependent. Table 9.1 contains descriptions of a few photometric quantities related to light intensity (these are subjective units), and table 9.2 contains details of a few radiometric quantities (absolute units).[1]

We will be using details of the terms given in table 9.2 later in this chapter, and in subsequent chapters. In radiometry, radiant flux, denoted Φ_e (subscript 'e', to avoid

[1] In photometry, there is an equivalent set of units to those in use in radiometry; however, photometry differs from radiometry in that quantities such as radiant flux are adjusted to reflect the varying sensitivity of the human eye to different wavelengths of light. Photometric quantities are denoted with a subscript 'v' (for 'visual') to avoid confusion with radiometric quantities, which are labelled with a subscript 'e' (for 'energy'). In photometry, luminous energy is the perceived energy of light. This is sometimes called the 'quantity of light.' However, luminous energy is not the same as radiant energy, the corresponding objective physical quantity. This is because the human eye can only see light in the visible spectrum, and has different sensitivities to light of different wavelengths within the spectrum. When adapted for bright conditions (photopic vision), the eye is most sensitive to light at a wavelength of 555 nm. Light with a given amount of radiant energy will have more luminous energy if the wavelength is 555 nm than if the wavelength is longer or shorter. Light whose wavelength is well outside the visible spectrum has a luminous energy of zero, regardless of the amount of radiant energy present.

Table 9.1. Some photometric quantities (in the SI).

Quantity	Symbol	Description	Unit	Dimension
Luminous energy	Q_v	The perceived energy of light.*(see footnote above)	Lumen second	JT
Luminous intensity	I_v	Luminous flux per solid angle	candela (cd) = lumen per steradian	J
Luminous flux	Φ_v	Luminous energy per unit time	lumen (= candela steradians) or lm = cd se	J
Luminance	L_v	Luminous flux per unit solid angle per unit projected source area	$cd/m^2 = lm/(Sr\ m^2)$	JL^{-2}

confusion with photometric units), is: $\Phi_e = \partial Q_e/\partial T$, where Q_e is the radiant energy emitted, reflected, transmitted or received, and T is time. From table 9.2, the radiant flux, Φ_e, is defined as the radiant energy emitted, reflected or received per unit time (that is, it is power with dimensions $ML^2T^{-2}/T = ML^2T^{-3}$). The corresponding radiant exitance (radiant emittance is an older name), M_e, is the radiant flux emitted by a surface per unit area (A). That is, $M_e = \partial\Phi_e/\partial A$, and dimensionally, $[M_e] = ML^2T^{-3}/L^2 = MT^{-3}$. Spectral flux can be defined in two ways—in terms of frequency of radiation (ν), or the wavelength of the radiation (λ). The spectral flux, in terms of frequency, $\Phi_{e,\nu}$ is defined as: $\partial\Phi_e/\partial\nu$, and the spectral flux in terms of wavelength, $\Phi_{e,\lambda}$ is defined as: $\partial\Phi_e/\partial\lambda$. We will use these quantities in the next chapter.

There is often confusion between photometric and radiometric quantities. This confusion arises because of the base dimension, J, used to define luminous intensity (see table 9.1). For example, the luminous intensity for light, of a particular wavelength λ is given by (see table 9.1):

$$I_v(\lambda) = 683\ y^*(\lambda) I_e(\lambda),$$

where $I_v(\lambda)$ is the luminous intensity in candelas (a photometric quantity), and $I_e(\lambda)$ is the radiant intensity in W/Sr (a radiometric quantity—see table 9.2), and $y^*(\lambda)$ is the spectral luminous efficiency, which represents the human eye's sensitivity to the light at a given wavelength. Hence, the candela is a photometric unit giving information about the perceived brightness of a source; in contrast, power, irradiance, radiant intensity, and radiance are radiometric units providing information about the absolute energy characteristics of a source.

The radiant intensity (power per solid angle), I_e has units of $W\ Sr^{-1} = J\ Sr^{-1}s^{-1}$, with $[I_e] = ML^2T^{-3}$, which is power. A lumen is a unit equivalent to a candela steradian; that is, lm = cd Sr, such that the luminous coefficient, approximately 683 lm/W = 683 cd Sr W^{-1}. Putting the units into the equation for the luminous intensity:

$$I_v(\lambda) = 683\ y^*(\lambda)\{cdSrW^{-1}\} I_e(\lambda)\{WSr^{-1}\}$$

Table 9.2. Some radiometric quantities (in the SI).

Quantity	Symbol	Description	Unit	Dimension
Radiant energy	Q_e	Energy radiated	joule	ML^2T^{-2}
Radiant power (flux)	Φ_e	Radiant energy emitted, reflected, transmitted or received, per unit time; that is, power, W = J/s, or $\partial Q_e/\partial T$	watt	ML^2T^{-3}
Radiant intensity	I_e	$\partial \Phi_e/\partial \Omega$ where Ω is the solid angle	watt/Sr	ML^2T^{-3}
Radiant exitance	M_e	Radiant flux emitted by a surface per unit area; $\partial \Phi_e/\partial A$	watt/m^2	MT^{-3}
Irradiance	E_e	Radiant flux received by a surface per unit area; $\partial \Phi_e/\partial A$	watt/m^2	MT^{-3}
Radiance	$L_{e,\Omega}$	Radiant flux emitted, reflected, transmitted or received by a surface, per unit solid angle per unit projected area; $\partial^2 \Phi_e/\partial A \partial \Omega$	watt/m^2 Sr	MT^{-3}
Spectral radiance	$(L_{e,\,\Omega,\,\nu})$	Radiance of a surface per unit frequency. $L_{e,\,\Omega,\,\nu} = \partial\, L_{e,\Omega}/\partial \nu$	Watt per steradian per square metre per hertz; W Sr^{-1} m^{-2} Hz^{-1}	MT^{-2}
Spectral radiance	$(L_{e,\,\Omega,\,\lambda})$	Radiance of a surface per unit wavelength. $L_{e,\,\Omega,\,\lambda} = \partial\, L_{e,\Omega}/\partial \lambda$	Watt per steradian per square metre per metre; W Sr^{-1} m^{-3}	$ML^{-1}T^{-3}$
Radiosity	J_e	Radiant flux leaving (emitted, reflected and transmitted by) a surface per unit area	Watt per square metre	MT^{-3}

(and cancelling units) we see that luminous intensity has units of candela, cd (and dimension of J). Indeed, this uniqueness of base dimension, J, is the reason many ask why is light intensity a base quantity in the SI? The definitions of the seven base quantities of the SI are defined in table 1.3; together with the level of precision of the value of the defining constant of Nature. Note the very different levels of precision associated with these quantities; the frequency of the hyperfine transition in a ^{133}Cs atom is quoted to ten significant figures, but the luminous efficacy, K_{cd}, is only quoted to three significant figures. With the creation of the Quantum-SI in 2019, and the coming redefinition of the second [2] in terms of an optical clock, rather than a microwave oscillator, there could well be moves by the international metrology community to reduce the number of base quantities in the SI from seven to six.

9.2 Polarizability

Imagine a sphere of radius, a. This sphere contains negative charge throughout its volume, but has positive charge at its centre to give electrical neutrality. We now apply an external electric field, $\boldsymbol{E}$, to this sphere. The effect of this applied field will be to polarize the sphere. The electrical charges in the sphere will separate, and the sphere will become elliptical, with its long axis parallel to the direction of the applied field. There will be a displacement, ξ, in charge distribution—the centre of the distribution of negative charge, will no longer be at the same point as the distribution of positive charge.

The new attractive force in the distorted sphere is proportional to $-e(\xi/a)^3$; where e is the electron charge, and we may write (in SI units):

$$\{e^2/(4\pi\varepsilon_0)\}a^3\boldsymbol{\xi} = -e\boldsymbol{E}. \tag{9.1}$$

This displacement of the point of balance of the charge distributions in the distorted sphere is an induced electric dipole moment, μ, and has dimensions of length and charge, $[\mu]$ = LTI, and is given by,

$$\mu = -e\boldsymbol{\xi} = (4\pi\varepsilon_0)a^3\boldsymbol{E}. \tag{9.2}$$

This is often written as $\mu = \varepsilon_0\alpha\boldsymbol{E}$, where $\alpha = 4\pi a^3$. This is the electric polarizability, which tells us how an atom, or a molecule, will respond to an applied electric field—either a static or an oscillating (optical) electric field. It is a property of all atoms, and in the c.g.s. system of units it has the dimension of volume, L^3. So, the polarizability of a helium atom (0.2×10^{-24} cm^3) is a lot smaller than that of a benzene molecule (9.96×10^{-24} cm^3), or a molecule of C_{60} (79×10^{-24} cm^3).

This analysis also tells us that an atom has a natural vibration, at a frequency ω_0; the attractive force associated with the displacement ξ being $m\omega_0{}^2\xi$. This gives,

$$\alpha = e^2/\varepsilon_0\, m\omega_0{}^2, \text{ or } \omega_0{}^2 = e^2/(4\pi\varepsilon_0)ma^3. \tag{9.3}$$

Here, m is the mass of the negatively-charged particle, which is very much smaller than the mass of the positively-charged particle, which leads to a natural frequency, of order, $\omega_0 \sim 10^{16}$ Hz. This is the origin of anomalous dispersion. The polarizability as a dynamical system can be investigated by dimensional analysis. We can look at the polarizability as the oscillation of a bound charged particle. The force acting on the mass can be calculated using Newton's second law, $F = ma = -m\omega^2 x$, where x is the amplitude of displacement, and ω is the natural resonant frequency defined for a mass, m in a potential described by F. This equation of force is similar to the force exerted by a spring (Hooke's law); that is, $F = -kx$. Comparing these last two equations we conclude that $k = m\omega^2$. Our dynamical model of the polarizability is equivalent to a harmonic oscillator whose frequency is given by $\omega = \sqrt{k/m}$, and whose period, t, is given by $t = 2\pi/\omega$.

Let us consider the amplitude of motion, x, of a charged particle (of charge e) bound by some force, with a force constant, k. That is: $x \propto (e)(\varepsilon_0)(force)$; where ε_0 is the permittivity, and from which we can set up the dimensional equation,

$$[x] = \mathsf{L} = (e)^a(\varepsilon_0)^b(\text{force})^c,$$

$$= (\mathsf{TI})^a(\mathsf{M}^{-1}\mathsf{L}^{-3}\mathsf{T}^4\mathsf{I}^2)^b(\mathsf{MLT}^{-2})^c = \mathsf{M}^{-b+c}\mathsf{L}^{-3b+c}\mathsf{T}^{a+4b-2c}\mathsf{I}^{a+2b}. \quad (9.4)$$

The solutions are: $a = 1$, $b = -½$ and $c = -½$. Thus, the amplitude of motion of the charged particle is

$$x = e/\sqrt{\varepsilon_0}\, k, \text{ or } x^2 = e^2/\varepsilon_0 k. \quad (9.5)$$

In a harmonic potential, the force constant, k, is given by the mass and the oscillation frequency of the particle in the bound potential; that is, $k = m\omega^2$. Thus, for the dynamic polarizability we may write $\alpha \sim e^2/\varepsilon_0\, m\omega^2$. This expression has dimensions: $[\alpha] = (\mathsf{TI})^2/\mathsf{M}^{-1}\mathsf{L}^{-3}\mathsf{T}^4\ \mathsf{I}^2\ \mathsf{MT}^{-2} = \mathsf{L}^3$; the c.g.s. dimension of the polarizability. The SI unit for the polarizability is $C^2m^2J^{-1}$.

9.2.1 Dimensional analysis applied to intermolecular forces

The electric field, V, has dimensions, $\mathsf{MLT}^{-3}\mathsf{I}^{-1}$, and in the SI has units of kg m s^{-3} A^{-1}. Here, one volt is defined as the difference in electric potential between two points of a conducting wire when an electric current of one ampere dissipates one watt of power between those points. It can also be expressed as: $V = A.\Omega = \frac{Wb}{s} = \frac{W}{A} = \frac{J}{C} = \frac{eV}{e}$.

The commonly used laboratory unit of volts per metre is thus, Vm^{-1} = kg $m^2s^{-3}A^{-1}$ m– = kg m $s^{-3}A^{-1}$, and an electric field gradient would be Vm^{-2} = kg $s^{-3}A^{-1}$. Dimensionally, we could start by saying that the electric field is proportional to the electric charge and distance; that is, $E \propto (e)(r)$. In fact, in the SI, we also have to consider the energy required to create the field in the vacuum, and so must also include the permittivity of free space, ε_0. Thus, $E \propto (e)(r)(\varepsilon_0)$. We know the dimensions of these variables, and so we can write:

$$[E] = (e)^a(r)^b(\varepsilon_0)^c. \quad (9.6)$$

The exponents are readily found from the dimensional equation,

$$\mathsf{MLT}^{-3}\mathsf{I}^{-1} = (\mathsf{TI})^a(\mathsf{L})^b(\mathsf{M}^{-1}\mathsf{L}^{-3}\mathsf{T}^4\mathsf{I}^2)^c = \mathsf{M}^{-c}\mathsf{L}^{b-3c}\mathsf{T}^{a+4c}\mathsf{I}^{a+2c}. \quad (9.7)$$

The fact that a occurs with c in the manner seen here, suggests that ε_0 is essential in explaining the phenomenon of electric phenomenon in the SI system of units. The exponents are: $a = 1$, $b = -1$ and $c = -1$. Thus, the electric field is represented by the equation $E = e/\varepsilon_0\, r$. Dimensional analysis has quickly revealed the details of the formation of an electric field, but we have lost all the numerical constants; for example, ε_0 rarely appears alone, but always with 4π. But such numerical factors, being dimensionless, are invisible to dimensional analysis.

The polarizability is the response of a molecule to an applied electric field. The potential energy of the molecule is lowered when the electric field is applied, and an electric dipole moment (μ) will be induced. This lowering of the potential energy in the presence of an applied electric field is why condensed phases exist. When molecules, particularly polarizable molecules, are brought together, the intermolecular forces can be quite significant, of order $k_B T$; and this attractive interaction leads to the formation of liquids and solids (covalently bonded materials). The dipole moment induced by the applied electric field is given by $\mu = \alpha E$; and so, the dimensions of the polarizability will be

$$[\alpha] = \frac{[\mu]}{[E]} = \mathsf{LTI/MLT^{-3}I^{-1}} = \mathsf{M^{-1}T^4I^2}. \tag{9.8}$$

In terms of units, C m/V m^{-1} = s.A m/kg m s^{-3} A^{-1} = kg^{-1} s^4 A^2.

As mentioned above, an electric field can be generated at a molecule by the presence of a neighbouring molecule; for example, the permanent dipole moment (which exists in a molecule because of charge asymmetry) can, like a bare electrical charge, generate an electric field. The only difference from the field due to a charge will be the radial dependence. Thus, the electric field of an electric dipole is:

$$E \propto (\mu)(r)(\varepsilon_0), \quad \text{where} \quad [E] = (\mu)^a(r)^b(\varepsilon_0)^c. \tag{9.9}$$

The dimensions are given in table 9.3. We may write the dimensional equation as

$$\mathsf{MLT^{-3}I^{-1}} = (\mathsf{LTI})^a(\mathsf{L})^b(\mathsf{M^{-1}L^{-3}T^4I^2})^c = \mathsf{M}^{-c}\mathsf{L}^{a+b-3c}\mathsf{T}^{a+4c}\mathsf{I}^{a+2c}. \tag{9.10}$$

The exponents are: $a = 1$, $b = -3$ and $c = -1$. Thus, the electric field generated by an electric dipole moment is $E = \mu/\varepsilon_0 r^3$. Given a typical value for a dipole moment (1 Debye $\approx 3.3 \times 10^{-30}$ Cm), the electric field at a distance of 1 Å is, of order, 10^{10} Vm^{-1}; which is much larger than any electric field it is possible to generate in

Table 9.3. Variables for investigating Rayleigh scattering and related phenomena (given in bold).

Variable	Dimension	Description
I	$\mathsf{MT^{-2}}$	Spectral radiance, ($L_{e,\ \Omega,\ \nu}$). Measured intensity of light scattered by an induced electric dipole (see table 9.2)
L	L	A length scale, which due to the nature of dimensional analysis will be the product of all length scales involved
μ	LTI	The dipole induced by the incident radiation, which then goes on to scatter that radiation
c	$\mathsf{LT^{-1}}$	Speed of light
ν	$\mathsf{T^{-1}}$	Frequency of incident light
μ_0	$\mathsf{MLT^{-2}\ I^{-2}}$	Permeability of free space
ε_0	$\mathsf{M^{-1}L^{-3}T^4\ I^2}$	**Permittivity of free space**
α	$\mathsf{M^{-1}T^4\ I^2}$	**Molecular polarizability**

the laboratory. The electric field generated by a permanent molecular quadrupole moment (Θ) in a non-dipolar molecule such as CO_2 or benzene would be $E = \Theta/\varepsilon_0 r^4$. Again, we are losing all the numerical factors by using dimensional analysis.

For a pair of neighbouring molecules, the electric field at molecule 1, due to the asymmetry in the charge distribution of molecule 2, which has a dipole moment, μ, could be written as $\boldsymbol{E}_{(1)} = \boldsymbol{E}_{(1)0} + \boldsymbol{\mu}_{(2)}\mathbf{T}$. Here we are only considering the contribution of the permanent electric dipole moment of molecule 2. The first term on the RHS is any static applied field at molecule 1. The second-rank tensor $\mathbf{T}$ is given by $(4\pi\varepsilon_0)$ $T_{\beta\gamma} = \nabla_\beta\nabla_\gamma(1/\boldsymbol{R}) = (3R_\beta R_\gamma - R^2\delta_{\alpha\beta})\ R^{-5}$ and $\boldsymbol{R}$ is the vector from molecule 1 to molecule 2; with $\nabla_\gamma(1/\boldsymbol{R}) = \partial(1/R)/\partial R_\beta = -R_\gamma R^{-3}$, and $\nabla_\alpha\nabla_\beta\nabla_\gamma(1/\boldsymbol{R}) \approx R^3/R^7 \approx R^{-4}$.[2]

In addition to a permanent dipole moment being able to exert an electric field on a neighbouring molecule, it is also possible for an induced dipole moment to exert a weak electric field. That is, a transient dipole moment is induced in a molecule or atom, and this weak dipole moment can then couple with the dipole moment (induced or permanent) of a neighbouring molecule to generate an attractive intermolecular potential. In the same way that an electric dipole can exert an electric field, two neighbouring dipoles can interact to generate an attractive (or a repulsive, depending upon the polarity of the two interacting dipoles) potential. This intermolecular potential, $E(r)$, between the electric dipole moments on two neighbouring molecules (1 and 2) will be $E(r) \propto (\mu_1)(\mu_2)(r)(\varepsilon_0)$, from which we may construct a dimensional equation,

$$[E] = (\mu_1)^a(\mu_2)^b(r)^c(\varepsilon_0)^d;\ \text{where,} \tag{9.11}$$

$$\mathsf{ML^2T^{-2}} = (\mathsf{LTI})^a(\mathsf{LTI})^b(\mathsf{L})^c(\mathsf{M^{-1}L^{-3}T^4I^2})^d = \mathsf{M}^{-d}\mathsf{L}^{a+b+c-3d}\mathsf{T}^{a+b+4d}\mathsf{I}^{a+b+2d}. \tag{9.12}$$

The solutions are: $a = b = 1$, $c = -3$ and $d = -1$. Thus, the intermolecular energy of interaction of two electric dipole moments is

$$E(r) = \mu_1\mu_2/\varepsilon_0 r^3. \tag{9.13}$$

A similar analysis of the interaction energy between two permanent electric quadrupole moments would begin with $E(r) \propto (\Theta_1)(\Theta_2)(r)(\varepsilon_0)$, and yields $E(r) = \Theta_1\Theta_2/\varepsilon_0 r^5$. And for the interaction energy of a dipole moment on one molecule and a quadrupole moment on another molecule, we would find $E(r) = \mu_1.\Theta_2/\varepsilon_0 r^4$. Again, all numerical factors (and they are becoming significant) are lost.

If we had been interested in determining the force between two electric dipoles; that is, $F(r)$ as opposed to $E(r)$ we would have started with

$$[F] = (\mu_1)^a(\mu_2)^b(r)^c(\varepsilon_0)^d;\ \text{where,} \tag{9.14}$$

$$\mathsf{MLT^{-2}} = (\mathsf{LTI})^a(\mathsf{LTI})^b(\mathsf{L})^c(\mathsf{M^{-1}L^{-3}T^4I^2})^d. \tag{9.15}$$

[2] I am employing standard Cartesian tensor notation in which a repeated Greek suffix denotes a sum over all three Cartesian components.

Only one of the dimensional equations would change from the analysis given above, L: $a + b + c - 3d = 1$, which would generate a different value for c, and give us the intermolecular force as $F(r) = \mu_1\mu_2/\varepsilon_0 r^4$. Thus, $F(r) = E(r)/\mathrm{L}$. If we were interested in the integrated energy of interaction, the power (P), we would have to introduce the speed of light, c (LT^{-1}), and would find $P = c\mu_1\mu_2/\varepsilon_0 r^4$.

Finally, let us look at a dimensional analysis of the interaction of an induced dipole moment with a permanent dipole; that is the dipole-induced dipole interaction. This interaction will be a modification of the intermolecular potential between two molecules, due to a dipole induced in one molecule through that molecule's polarizability, and so is also an energy contributing to the intermolecular potential, $E(r)$. Thus, we start by identifying what we consider to be the essential variables in a model of the changing character of the intermolecular potential due to this weak electrostatic-interaction: $E(r) \propto (\alpha)^a(\mu)^b(r)^c(\varepsilon_0)^d$. Here, α is the polarizability of one molecule, and the dipole moment is on a neighbouring molecule. The dimensional equation is (the dimensions are given in table 9.2):

$$\mathsf{ML^2T^{-2}} = \left(\mathsf{M^{-1}T^4I^2}\right)^a(\mathsf{LTI})^b(\mathsf{L})^c\left(\mathsf{M^{-1}L^{-3}T^4I^2}\right)^d = \mathsf{M}^{-a-d}\mathsf{L}^{b+c-3d}\mathsf{T}^{4a+b+4d}\mathsf{I}^{2a+b+2d}. \quad (9.16)$$

The solutions are: $a = 1$, $b = 2$, $c = -6$ and $d = -2$. Thus,

$$E(r) = \alpha\mu^2/r^6\varepsilon_0^2. \quad (9.17)$$

Given that the permittivity appears squared, and that it occurs as $4\pi\varepsilon_0$ in the full theory, we get a feeling for the magnitude of the numerical factors that are being lost by the use of dimensional analysis. What this analysis tells us is that the term $\alpha\mu^2/r^6\varepsilon_0^2$ has the units of energy. This is easily checked: $\alpha\mu^2/r^6\varepsilon_0^2 = \mathrm{C^2\ m^2\ J^{-1}\ C^2\ m^2/F^2\ m^{-2}\ m^6} = \mathrm{C^4\ m^4\ J^{-1}\ m^{-6}/kg^{-1}\ m^{-3}\ s^4\ A^2} = \mathrm{kg\ m^2\ s^{-2}} = $ Energy.

This analysis of the interaction between the electric field of a dipole moment with that of a transient dipole induced in a neighbouring molecule tells us a great deal about molecular interactions. First, molecule 1 is polarized by the electric field arising from a pre-existing electric moment on a neighbouring molecule, 2; that is, a dipole is induced on molecule 1. This coupling of the field from molecule 2 to the polarizability of molecule 1 can be represented by the **T**-tensors described above.

$$\mu_{(1)} = \alpha_{(1)}.E_{(2)} = \alpha_{(1)}.\mathbf{T}_{(1-2)}\,\mu_{(2)}/(4\pi\varepsilon_0) \approx \alpha_{(1)}\,\mu_{(2)}/(4\pi\varepsilon_0)r^3. \quad (9.18)$$

This gives us two dipole moments—one larger in magnitude than the other. These two dipoles will generate an attractive intermolecular potential,

$$E_{(1-2)} = \mu_{(1)}.\,\mu_{(2)}/(4\pi\varepsilon_0)r^3. \quad (9.19)$$

Substituting equation (9.18) into equation (9.19) yields

$$E_{(1-2)} = \left\{\alpha_{(1)}\,\mu_{(2)}/(4\pi\varepsilon_0)r^3\right\}\mu_{(2)}/(4\pi\varepsilon_0)r^3 = \left\{\alpha_{(1)}\,\mu_{(2)}^{\ 2}/r^6\right\}1/(4\pi\varepsilon_0)^2. \quad (9.20)$$

We see here, that we divide twice by $4\pi\varepsilon_0$; firstly, to induce the dipole moment, and secondly to consider the energy of interaction between the two dipole moments. This also means that this contribution varies as $1/r^6$ of the separation of the two molecules; consequently, it is always additive and attractive. In this, it is just like the closely-related London-dispersion force.

9.3 Rayleigh scattering

The investigation of the interaction of light with matter has a long history. In 1871, Lord Rayleigh treated light as a mechanical vibration in the luminiferous aether [3]. Rayleigh discovered that the intensity of light scattered by particles was much smaller than the wavelength (λ) of the incident light but to be proportional to $1/\lambda^4$. In 1881, Rayleigh deduced the same result using James Clerk Maxwell's electromagnetic wave theory [3]. Thus, Lord Rayleigh explained why the sky is blue; as incident white light from the Sun is scattered by the molecules that compose the atmosphere, the shortest wavelengths contribute most ($1/\lambda^4$) to the scattering (see figure 9.2).

Rayleigh scattering is the scattering of light by induced electric dipoles. Consider an electric dipole, μ, induced in a molecule by an oscillating electric field (a light-wave). The applied electric field is $E = E_0\, e^{i\omega t}$, and the induced dipole moment is $\mu = \alpha\, E = E_0\, e^{i\omega t}$, where α is the polarizability of the molecule. The radiation field of this induced, oscillating dipole moment will be,

$$\mathbf{E} = -k_0{}^2\alpha E_0 \sin\theta(\mathrm{e}^{\mathrm{i}(\omega t - x\mathrm{k}0r)})/r\ \mathbf{i}_2, \quad \text{and } \mathbf{H} = \chi E\ \mathbf{i}_3. \tag{9.21}$$

Thus, the time average of the Poynting vector (in c.g.s. units) is: $I_0 = \frac{c}{8\pi}\,|E_0|^2$; and $I = \frac{c}{8\pi}\,|E|^2$, and $I_0/I = k_0{}^4\ \alpha^2\ \sin^2\theta/r^2$, where $k_0 = \omega/c = 2\pi/\lambda$, and I and I_0 are,

Figure 9.2. The blue colour of the sky is caused by light scattered by atmospheric gas particles, which are much smaller than the wavelengths of visible light; the red colour at sunset and sunrise comes from absorption. The grey/white colour of clouds is caused by light scattered by water droplets, which are of a comparable size to the wavelengths of the incident visible light. Visible light has a wavelength of, typically, 5400 Å in the green, whereas small molecules have a 'diameter' of about 1–2 Å.

respectively, the scattered and the incident intensities. With N molecules at ambient pressure,

$$I/I_0 = N\, k_0^4\, \alpha^2 \sin^2\theta/r^2. \tag{9.22}$$

The scattering is observed along an axis at a distance, R from the viewer. Integrating the scattered intensity over a full sphere, we find the total energy loss per unit time:

$$\int_0^{\pi} I\, 2\pi\, r^2 \sin\theta\, d\theta = 2\pi N\, k_0^4\, \alpha^2\, I_0 \int_{-1}^{1} (1 - x^2)\, \mathrm{d}x = \frac{8\pi}{3} N\, k_0^4\, \alpha^2\, I_0. \tag{9.23}$$

The size of a scattering particle is often parameterized by the dimensionless ratio, $x = 2\pi r/\lambda$, where r is the radius of the scattering particle, λ is the wavelength of the incident light. This ratio characterizes the interaction of the particle with the incident radiation such that: Objects with $x \gg 1$ act as geometric shapes, scattering light according to their projected area. At the intermediate $x \simeq 1$ of Mie scattering, interference effects develop through phase variations over the object's surface. Rayleigh scattering applies to the case when the scattering particle is very small ($x \ll 1$) and the whole surface re-radiates with the same phase. Because the particles are randomly positioned, the scattered light arrives at a particular point with a random collection of phases; it is incoherent and the resulting intensity is just the sum of the squares of the amplitudes from each particle, and therefore proportional to the inverse fourth power of the wavelength and the sixth power of its size. The wavelength dependence is characteristic of dipole scattering.

Let us now look at a dimensional analysis of this phenomenon. Consider an incident electromagnetic wave with uniform amplitude E_i incident upon a particle, capable of scattering; that is, a polarizable particle of volume, $V \sim L^3$. If the wavelength of the light is large such that $x \ll 1$, where x is defined above, then the entire volume V will oscillate in phase, and hence the scattered amplitude, E_s, will be proportional to V, so $E_s \propto L^3$. That is, in the c.g.s. system of units, the polarizable volume defines the polarizability of the molecule—the bigger the molecule, the bigger the polarizability, and the greater the amount of light scattered. In contrast to the incident plane wave, the scattered wave is emitted in all directions, and hence the scattered intensity, I_s (which is proportional to the square of the amplitude), must fall off as the inverse square of the distance r from the scatterer, $I_s \propto E_s^2 \sim 1/r^2$. Finally, the scattered amplitude must also be proportional to the incident amplitude, $E_s \propto E_i$. Putting this all together leaves us with the wrong dimensions of length, and the only parameter with dimensions of length is the wavelength of the incident light beam, so the final version of the expression must be

$$I_s \propto E_s^2 \sim E_i^2(L^6/r^2\, \lambda^4) \propto \nu^4, \tag{9.24}$$

where the last proportionality arises because $c = \nu\, \lambda$, ν being the frequency of the radiation. Hence, we obtain the prominent feature of Rayleigh scattering.

Let us now look at a full Buckingham Π-theorem examination of Rayleigh scattering. We first set out the variables we believe to be of importance in explaining the phenomenon of Rayleigh scattering. Thus, we write

$$\Pi = (I)(L)(\mu)(c)(\nu)(\mu_0). \tag{9.25}$$

Here, I is the observed scattered light intensity, L is a length scale, μ is the induced electric dipole moment that then goes on to scatter radiation, c is the speed of light, ν is the frequency of the light scattered, and μ_0 is the permeability of the vacuum. These variables, together with their dimensions, are set out in table 9.3.

There are six variables not in bold, and these six variables involve four base dimensions. Thus, the Buckingham Π-theorem tells us that there are 6–4 = 2 nondimensional combinations of these variables to be identified. The next stage is to identify which variables will be the repeat variables, and which two variables will be the lead terms in constructing the two nondimensional combinations. Here, we chose to use the intensity to construct Π_1, and the frequency to construct Π_2. So, the repeat variables are: $(L)(\mu)(c)(\mu_0)$.

Thus, for Π_1 the dimensional equation is:

$$\Pi_1 = (I)(L)^a(\mu)^b(c)^c(\mu_0)^d = (\mathsf{MT}^{-2})\ (\mathsf{L})^a\ (\mathsf{LTI})^b(\mathsf{LT}^{-1})^c(\mathsf{MLT}^{-2}\mathsf{I}^{-2})^d. \tag{9.26}$$

From this expression we can derive four equations. For this, we have taken the scattered light intensity to be in watt persteradian per square metre per hertz; that is, W $\mathrm{Sr}^{-1}\mathrm{m}^{-2}\ \mathrm{Hz}^{-1}$, or MT^{-2} (see table 9.3: ($L_{e,\ \Omega,\ \nu}$)). Thus:

$$\begin{aligned}
&\mathsf{M}: 1 + d = 0\\
&\mathsf{L}: a + b + c + d = 0\\
&\mathsf{T}: -2 + b - c - 2d = 0\\
&\mathsf{I}: b - 2d = 0.
\end{aligned}$$

This gives: $b = -2$, $a = 5$, $c = -2$ and $d = -1$; and

$$\Pi_1 = (I)(L)^5(\mu)^{-2}(c)^{-2}(\mu_0)^{-1} = IL^5/\mu^2\mu_0\, c^2. \tag{9.27}$$

The second nondimensional combination of variables is found in an analogous manner. Thus,

$$\begin{aligned}
\Pi_2 &= (L)^a(\mu)^b(c)^c(\mu_0)^d(\nu)\\
&= (\mathsf{L})^a(\mathsf{LTI})^b(\mathsf{LT}^{-1})^c(\mathsf{MLT}^{-2}\mathsf{I}^{-2})^d(\mathsf{T}^{-1}).
\end{aligned} \tag{9.28}$$

This time, we have:

$$\begin{aligned}
&\mathsf{M}: d = 0\\
&\mathsf{L}: a + b + c + d = 0\\
&\mathsf{T}: -1 + b - c - 2d = 0\\
&\mathsf{I}: b - 2d = 0.
\end{aligned}$$

We see that $d = 0$, which means that $b = 0$; the other terms are: $a = 1$ and $c = -1$. Thus, the second nondimensional group is

$$\Pi_2 = L\nu / c. \tag{9.29}$$

The Buckingham Π-theorem allows us to relate these two nondimensional groups as

$\Pi_1 = f(\Pi_2)$. Thus,

$$IL^5/\mu^2 \mu_0 c^2 = f\left(\frac{Lv}{c}\right), \tag{9.30}$$

and

$$I = \mu^2 \mu_0 c^2 L\omega / cL^5 = \mu^2 \mu_0 cv / L^4. \tag{9.31}$$

Even without the significant numerical factors lost to the dimensional analysis, we can see that a page of dimensional analysis has given us the essentials of Rayleigh scattering; particularly, that it is dipole radiation (induced-dipole dipole scattering) and the famous $1/\lambda^4$ dependence of wavelength. So, blue-light is scattered more strongly than red-light (by a factor, of order, 5), and the sky appears blue when the Sun is at its zenith.

9.4 Collision-induced light scattering

As another example of the use of dimensional analysis in a complex problem in chemical physics, we will look briefly at the modelling of the intensity of light scattered by centrosymmetric molecules such as carbon dioxide, and carbon disulphide in the liquid state. As is well known, in linear centrosymmetric molecules the bending (Π_u) and asymmetric stretching (Σ_u) vibrational modes are infrared-active and Raman-forbidden in the isolated molecule. However, it is a well-established observation that in the liquid, or compressed gas, the electric dipole selection rules break down, and the bending and asymmetric stretch become weakly allowed in the Raman spectrum. As with other forms of collision-induced spectra, this breakdown of the selection rules may be attributed to intermolecular interactions, and, in particular to the effects of the electric field, and the gradient of the electric field produced at one molecule by neighbouring molecules.

Two mechanisms were proposed to explain these collision-induced vibrational spectra in CO_2 [4]. One involved fluctuations in the first hyperpolarizability, $\boldsymbol{\beta}$, leading to an oscillating dipole moment in the presence of the optical electric field $\boldsymbol{E}_{(o)}$ of the probing laser and the strong field due to the charge distribution of a neighbouring molecule. The second mechanism depended upon fluctuations in the dipole-quadrupole polarizability, $\boldsymbol{A}$, through which a dipole is induced by the field gradient due to the dipole already induced in a neighbouring molecule. It was demonstrated [4] by detailed analysis that it was the latter mechanism that best accounts for the observed collision-induced light scattering in liquid CO_2.

Consider two separate, but neighbouring molecules in an optical field $\boldsymbol{E}_{(0)}$. The total electric field $\boldsymbol{E}_{(1)}$ at molecule 1 is the sum of $\boldsymbol{E}_{(0)}$ and the field $\boldsymbol{E}_{(12)}$, which arises from the charge distribution of molecule 2. The dipole moment, μ, of molecule 1 may then be expressed as

$$\mu_{(1)\alpha} = \mu_{(0)(1)\alpha} + \alpha_{(1)\alpha\beta}E_{(12)\beta} + 1/2\beta_{(1)\alpha\beta\gamma}E_{(12)\beta}E_{(12)\gamma} + 1/3A_{(1)\alpha\beta\gamma}E_{(12)\beta\gamma} + \dots \quad (9.32)$$

Here, $\mu_{(0)(1)}$ is any permanent dipole moment on molecule 1, $E_{(12)\beta}$ is the electric field at molecule 1 due to charge asymmetry in molecule 2, and $E_{(12)\beta\gamma}$ is a component of the electric field gradient at molecule 1 due to the dipole ($\boldsymbol{\mu}$) and quadrupole ($\boldsymbol{\Theta}$) moments of neighbouring molecule 2; for example,

$$E_{(12)\beta} = T_{\beta\gamma}\mu^{(2)}{}_{\gamma} + 1/3T_{\beta\gamma\delta}\Theta^{(2)}{}_{\gamma\delta} + \dots, \quad (9.33)$$

and

$$E_{(12)\beta\gamma} = -T_{\beta\gamma\delta}\mu^{(2)}{}_{\delta-1/3}T_{\beta\gamma\delta\varepsilon}\Theta^{(2)}{}_{\delta\varepsilon} + \dots, \quad (9.34)$$

where the interaction tensors, $\mathbf{T}$, are defined above.

The effective polarizability, which is what generates the collision-induced Raman spectrum, of the pair of molecules is:

$$\pi_{\alpha\beta} = \pi_{(1)\alpha\beta} + \pi_{(2)\alpha\beta} = \left(\partial\mu_{(1)\alpha}/\partial E_{(0)\beta}\right) + \left(\partial\mu_{(2)\alpha}/\partial E_{(0)\beta}\right), \quad (9.35)$$

and the intensity of the Raman spectrum depends upon the derivative of $\pi_{\alpha\beta}$ with respect to the particular normal coordinate q. We assume that $q = q_{(1)}$ is the normal coordinate of a u-mode of vibration of molecular 1: $\partial\boldsymbol{\alpha}/\partial q$, $\boldsymbol{\beta}$ and $\boldsymbol{A}$ vanish at $q = 0$ for centrosymmetric molecules. From equations (9.32) and (9.35) we have

$$\begin{aligned} \pi'_{\alpha\beta} &= \left(\partial\pi_{\alpha\beta}/\partial q\right) = \left(\partial\beta_{(1)\alpha\beta\gamma}/\partial q\right)E_{(12)\gamma} \\ &\quad + \tfrac{1}{3}\left(\partial A_{(1)\alpha\gamma\delta}/\partial q\right)\left(\partial E_{(12)\gamma\delta}/\partial E_{(0)\beta}\right) + \alpha_{(2)\alpha\gamma}\left(\partial^2 E_{(21)}\gamma/\partial q\partial E_{(0)\beta}\right) \\ &= \tfrac{1}{3}\left(\partial\beta_{(1)\alpha\beta\gamma}/\partial q\right)T_{\gamma\delta\varepsilon}\Theta_{(2)\delta\varepsilon} - \tfrac{1}{3}\left(\partial A_{(1)\alpha\gamma\delta}/\partial q\right)T_{\gamma\delta\varepsilon}\alpha_{(2)\varepsilon\beta} \\ &\quad - \tfrac{1}{3}\left(\partial A_{(1)\beta\delta\varepsilon}/\partial q\right)T_{\gamma\delta\varepsilon}\alpha_{(2)\alpha\gamma} - \dots = \pi'_{\beta\alpha}. \end{aligned} \quad (9.36)$$

The Raman intensity is proportional to the square of $\pi'_{\alpha\beta}$ averaged over all possible orientations. This complicated expression represents the interaction of only two molecules among the 10^{23} molecules that exist in an experimental system, and the expression is therefore naturally located within a molecule-fixed coordinate system. Experiments, however, are undertaken in a laboratory-fixed coordinate system; that is, a laser is polarized in a particular plane and the beam propagates in a particular direction within a laboratory-fixed axis system. Consequently, to compare this theoretical expression in the molecule-fixed coordinate system to an experimentally observed Raman intensity requires that equation (9.36) be transformed into the laboratory-fixed coordinate system. And this is done by taking an isotropic average of equation (9.36). That is, we require only the scalar quantities $\pi'_{\alpha\alpha}\,\pi'_{\beta\beta}$ and $\pi'_{\alpha\beta}\,\pi'_{\alpha\beta}$. Of the two molecular terms contributing to the induced intensity in equation (9.36), $(\partial\beta/\partial q)^{2}\Theta^2R^{-8}$ and $(\partial A/\partial q)^{2}\alpha^2R^{-8}$, the latter was shown to be the largest contribution [4].

We can also look at these complicated expressions dimensionally, and seek to check that no term has been overlooked, and also to see how much simpler is a dimensional analysis of this problem. The light is scattered by the differential polarizability, $\pi_{\alpha\beta} = (\partial\mu_\alpha/\partial E_\beta)$, the dimensions of this quantity may be found from those of the electric dipole moment, $[\mu]$ = LTI, and the dimensions of the electric field, $[E]$ = $\mathsf{MLT^{-3}I^{-1}}$. (Do not confuse the symbol for this differential polarizability with the Π of the nondimensional groups derived from the Buckingham Π-theorem.) Thus, LIT/$\mathsf{MLT^{-3}I^{-1}} = [\pi] = \mathsf{M^{-1}T^4I^2}$; these are also the dimensions of the electric polarizability, α, which is responsible for Rayleigh scattering and Raman scattering; $[\alpha] = \mathsf{M^{-1}T^4I^2}$. Table 9.4 contains the details of the dimensions involved in this problem.

We begin by asking, which variables might be involved in the collision-induced light scattering. Here, we suggest the molecular hyperpolarizability, $\boldsymbol{\beta}$, the molecular quadrupole moment, Θ, the distance between molecules, r, and the permittivity of free space, ε_0. We thus hypothesise that the collision-induced scattering, induced via the hyperpolarizability, is ($\pi^{(\beta)}$),

$$\pi^{(\beta)} = (\partial\mu/\partial E) = (\beta)^a(\Theta)^b(r)^c(\varepsilon_0)^d. \tag{9.37}$$

The dimensions for these molecular properties and constants may be found in table 9.3. The dimensional equation to help us define how the variables may be combined to give a combination of parameters with the dimensions of molecular polarizability (via the Rayleigh algorithm) is

$$[\pi^{(\beta)}] = [\alpha] = \mathsf{M^{-1}T^4I^2} = (\mathsf{M^{-2}L^{-1}T^7I^3})^a(\mathsf{L^2TI})^b(\mathsf{L})^c(\mathsf{M^{-1}L^{-3}T^4I^2})^d. \tag{9.38}$$

To calculate the exponents, we set up four linear equations, one for each dimension:

$$\begin{aligned}
&\mathsf{M}: -2a—d = -1\\
&\mathsf{L}: -a + 2b + c - 3d = 0\\
&\mathsf{T}: 7a + b + 4d = 4\\
&\mathsf{I}: 3a + b + 2d = 2.
\end{aligned}$$

Table 9.4. Variables for collision-induced light scattering.

Variable	Dimension	Description
π	$\mathsf{M^{-1}T^4I^2}$	Differential polarizability, which may be induced via the hyperpolarizability, β, or the dipole-quadrupole, A, polarizability (π^β, or π^A, respectively)
α	$\mathsf{M^{-1}T^4I^2}$	Polarizability
β	$\mathsf{M^{-2}\,L^{-1}\,T^7I^3}$	Hyperpolarizability
A	$\mathsf{M^{-1}\,L\,\,T^4I^2}$	Dipole-quadrupole polarizability
Θ	$\mathsf{L^2TI}$	Molecular quadrupole moment
r	L	Length coordinate
ε_0	$\mathsf{M^{-1}\,L^{-3}\,T^4\,I^2}$	Permittivity of free space

The solutions are: $a = 1$, $b = 1$, $c = -4$ and $d = -1$. These exponents tell us that the scattered intensity arising from the coupling of the molecular hyperpolarizability on one molecule with the electric field due to the quadrupole moment on a neighbouring molecule (at a separation, r) is

$$\pi^{(\beta)} = \Theta\beta/\varepsilon_0 r^4. \tag{9.39}$$

Here, the observed intensity, $I_{\text{obs}} \propto |<\pi^{(\beta)}>|^2$, where the angle brackets denote an averaging of the molecular properties over all angles. Thus,

$$I_{\text{obs}} \approx \Theta^2\beta^2/\varepsilon_0{}^2 r^8. \tag{9.40}$$

A similar analysis for the contribution to the observed intensity arising through the electric field gradient on a molecule acting upon the dipole-quadrupole polarizability, $\boldsymbol{A}$, is readily made. Here we start with a different assumption,

$$\pi^{(A)} = (\partial\mu/\partial E) = (A)^a(r)^b(\alpha)^c(\varepsilon_0)^d; \tag{9.41}$$

that is, the electric field gradient arising from a pre-existing, transient electric dipole moment induces a dipole moment, in the target molecule via the dipole-quadrupole polarizability, $\boldsymbol{A}$. The dipole-quadrupole polarizability describes the dipole induced by a field gradient or the quadrupole induced by an electric field. To determine the contribution to the measured scattered light from the dipole-quadrupole polarizability, we need the following dimensional equation (see table 9.4),

$$[\pi^{(A)}] = [\alpha] = \mathsf{M}^{-1}\mathsf{T}^4\mathsf{I}^2 = (\mathsf{M}^{-1}\mathsf{L}\mathsf{T}^4\mathsf{I}^2)^a(\mathsf{L})^b(\mathsf{M}^{-1}\mathsf{T}^4\mathsf{I}^2)^c(\mathsf{M}^{-1}\mathsf{L}^{-3}\mathsf{T}^4\mathsf{I}^2)^d. \tag{9.42}$$

A solution for the exponents arising in the four linear equations generated by this expression yields the following expression for A-tensor contribution to the differential polarizability,

$$\pi^{(A)} = \alpha A/\varepsilon_0 r^4. \tag{9.43}$$

Similarly, the observed intensity, $I_{\text{obs}} \propto |< \pi^{(A)} >|^2$, where the angle brackets denote an averaging of the molecular properties over all angles. Thus,

$$I_{\text{obs}} \approx \alpha^2 A^2/\varepsilon_0{}^2 r^8. \tag{9.44}$$

So, there are two contributions to the observed collision-induced light scattering in a condensed phase such as liquid CO_2 or CS_2. These two contributions are (variables described in table 9.3): via the hyperpolarizability, $\boldsymbol{\beta}$: $I_{\text{obs}} \approx \Theta^2\beta^2/\varepsilon_0{}^2r^8$; and via the dipole-quadrupole polarizability, $\boldsymbol{A}$: $I_{\text{obs}} \approx \alpha^2A^2/\varepsilon_0{}^2r^8$. By deriving these expressions via the mechanism of dimensional analysis, we have ensured that they have the same dimensions: $[\Theta^2\beta^2/\varepsilon_0{}^2r^8] = [\alpha^2A^2/\varepsilon_0{}^2r^8] = \mathsf{M}^{-2}\mathsf{T}^8\mathsf{I}^4$.

All numerical factors are eliminated by the dimensional analysis (and they are significant [4]). But the essential thing revealed by the dimensional analysis is that in a few lines we have duplicated the result of the extended analysis given in the full treatment of this problem [4]. Both terms capable of generating collision-induced light scattering have the dimensions of α^2. Thereby showing the similarity of this phenomenon with that of Rayleigh scattering.

References

[1] The SI Brochure, or The International System of Units (SI) 2019 https://www.bipm.org/en/publications/si-brochure/

[2] Williams J.H. 2020 *Defining and Measuring Nature: The Make of All Things* 2nd edn (Bristol: IOP Publishing) A full description of the background to the recent changes in the SI can be found here: Stock M, Davis R, de Mirandés E and Milton M J T 2019 The revision of the SI—the result of three decades of progress in metrology *Metrologia* **56**, 022001

[3] Lord Rayleigh 1899 On the transmission of light through an atmosphere containing small particles in suspension, and on the origin of the blue of the sky *Philos. Mag. J. Sci* **47** 375–384
Strutt Hon J W 1871 On the light from the sky, its polarization and colour *Philos. Mag. J. Sci.* **41** 107–120
Strutt Hon J W 1871 On the light from the sky, its polarization and colour *Philos. Mag. J. Sci.* **41** 274–9
Strutt Hon J W 1871 On the scattering of light by small particles *Philos. Mag. J. Sci.* **41** 447–54

[4] Amos R D, Buckingham A D and Williams J H 1980 Theoretical studies of the collision-induced Raman spectrum of carbon dioxide *Mol. Phys.* **39** 1519-1526

IOP Publishing

Dimensional Analysis
The great principle of similitude
Jeffrey H Williams

Chapter 10

The equilibrium between matter and energy

An experiment is a question which science poses to Nature, and a measurement is the recording of Nature's answer.

Max Planck (1858–1947)

One of the great problems of physics in the latter-half of the 19th Century was to identify the relationship between matter and energy. This endeavour would eventually result in the creation of quantum mechanics in 1900, and lead to the triumphs of the 20th Century. But back in 1864, the Irish physicist, John Tyndall (1820–1893) could only publish results of his experimental investigations of the radiation emitted by a hot metallic body (a filament). There was no theoretical framework within which to view these measurements—they were merely observations. However, an Austrian physicist examined the published data, and deduced the proportional relationship between the intensity of the emitted radiation, and the temperature of the filament. It was found that the radiated intensity, I_{radiated}, was proportional to the fourth-power of the temperature of the emitting filament; that is, $I_{\text{radiated}} \propto T^4$. This proportionality was deduced in 1879 by Josef Stefan (1835–1893). The derivation of a law, from theoretical considerations relating the intensity of emitted radiation and filament temperature was, subsequently, presented by Ludwig Boltzmann (1844–1906) in 1884; this theoretical result was quickly verified experimentally. In 1888, Heinrich Weber pointed out deviations at higher temperatures, but by 1897 reproducible measurements had been observed up to filament temperatures of 1525 K. The law, which we now call the Stefan–Boltzmann law, including the theoretical prediction of the Stefan–Boltzmann constant in terms of the fundamental constants of Nature, was finally formulated after the publication of the radiation law of Max Planck in 1900.

The best place to look at the advance of 19th Century physics, in rationalising the equilibrium between matter and energy is in a black-body; that is, a well-characterised object that emits radiation when heated. We are familiar with the principle as a heated iron-bar glows red-hot, and becomes white-hot when heated further. The metal-bar is emitting electromagnetic radiation; not only photons of wavelengths in the visible part of the spectrum, but also infrared and ultraviolet light. The distribution of wavelengths emitted by the black-body is a function of temperature. Figure 10.1 gives the concept of the measurement. A hollow sphere is held at a temperature, and the wavelength of the radiation emitted from the interior of the body, through a well-collimated hole, is observed and characterised.

Consider a black-body in thermal equilibrium with radiation. We assume that: the radiation energy which the black-body absorbs is converted into thermal energy only; $E_\nu d\nu$ denotes the amount of energy emitted by the black-body in unit time per unit area in the frequency interval $d\nu$, and A_ν is its absorption coefficient for frequency ν. In 1860, the German physicist Gustav Kirchhoff (1824–1887) proposed [1] a theorem that E_ν/A_ν depends only upon ν (the frequency of the radiation) and the temperature, T, and is independent of any other characteristic of the body; that is,

$$E_\nu/A_\nu = B(\nu, T), \tag{10.1}$$

where $B(\nu,T)$ is the exitance of the black-body (flux emitted by a surface per unit area). Gustav Kirchhoff never got to the bottom of the nature of this radiation, however, his proposal set in motion a series of experimental and theoretical developments that culminated in the creation of quantum mechanics. Kirchhoff defined a body for which $A_\nu = 1$; today we call this a perfect black-body. Then $B(\nu,T)$ is the radiant exitance of the black-body; today, labelled M_e (see table 9.2). Kirchhoff defined the perfect black-body: '*Given a space enclosed by bodies of equal temperature, through which no radiation can penetrate, then every bundle of radiation* [the classical

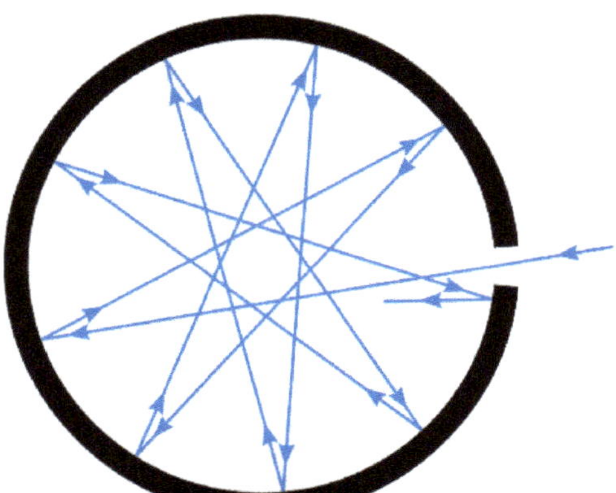

Figure 10.1. A conceptualised black-body emitting radiation. A small hole entrance to a large cavity with blackened walls. Any light entering through the hole is entirely absorbed. Since the hole is very small, the interior of the cavity contains radiation almost in thermal equilibrium with the walls maintained at a fixed temperature. Hence, the hole emits black-body radiation. This 'black body realization' image has been obtained by the author from the Wikimedia website where it was made available under a CC BY-SA 4.0 license. It is included within this article on that basis. It is attributed to AG Caesar.

way of describing a quantity of radiation] *within this space is constituted, with respect to quality* [frequency] *and intensity as if it were from a black-body of the same temperature.*' Kirchhoff challenged theorists and experimentalists alike with the '*highly important task*' of determining the function $B(\nu,T)$. Kirchhoff derived equation (10.1) by saying that its violation would imply the possibility of a perpetual motion machine of the second kind; that is, he defined it from the idea of the conservation of energy. At this time, thermodynamics was a new field of research and Kirchhoff's ideas were at the very forefront of theoretical physics. And not surprisingly, a quarter of a century was to pass before the next major contribution to this problem.

In 1879, Joseph Stefan conjectured, on experimental grounds, that the total energy radiated by a hot body varies with the fourth-power of the absolute temperature. This idea is only approximately true, and the precise formulation was given in 1884 when Boltzmann demonstrated that the strict T^4 dependence only holds for true black-body radiators. Boltzmann's proof is again dependent upon thermodynamics, but combined this time with an even newer branch of physics—Maxwell's electromagnetic theory. The problem put forward by Kirchhoff was becoming the nexus for the main strands of classical 19th Century physics.

The modern statement of the Stefan–Boltzmann law describes the power radiated from a black-body in terms of its temperature. Specifically, the Stefan–Boltzmann law says that the total energy radiated per unit surface area of a black-body across all wavelengths per unit time, M_e (also known as the black-body radiant exitance, see table 9.2) is directly proportional to the fourth-power of the black-body's thermodynamic temperature, T:

$$M_e = \sigma T^4. \tag{10.2}$$

The constant of proportionality, σ, is called the Stefan–Boltzmann constant; the value of the constant is today: $\sigma = \frac{2}{15}(\pi^5 k_B{}^4/c^2 h^3) = 5.670\,374\,419 \times 10^{-8}\ \mathrm{W\,m^{-2}\,K^{-4}}$, where k_B is the Boltzmann constant, h is the Planck constant and c is the speed of light. The value of the Stefan–Boltzmann constant is known with great precision. In fact, there is no uncertainty associated with this value as it is defined by three constants of Nature (c, h and k_B), whose values have been fixed by the 2019 creation of the Quantum-SI [2].

From table 9.2, the radiant flux, Φ_e, is defined as the radiant energy emitted, reflected or received per unit time (that is, it is power with dimensions $\mathsf{ML^2T^{-2}/T} = \mathsf{ML^2T^{-3}}$). The corresponding radiant exitance (radiant emittance is an older name), M_e, is the radiant flux emitted by a surface per unit area (A). That is, $M_e = \partial\Phi_e/\partial A$, and dimensionally, $[M_e] = \mathsf{ML^2T^{-3}/L^2} = \mathsf{MT^{-3}}$. The radiant exitance of a black-body, $M_e{}^\circ$ is an ideal concept, and was found by Stefan to be proportional to the fourth-power of the temperature of the black-body; $M_e{}^\circ \propto T^4 = \sigma T^4$. For a real emitting surface, other than a black-body surface, the radiant exitance is: $M_e = \varepsilon M_e{}^\circ = \varepsilon\sigma T^4$, where ε is the emissivity of that surface, which is its effectiveness in emitting energy as thermal radiation, which may include both visible radiation and infrared radiation. Quantitatively, emissivity is the ratio of the thermal radiation from a surface to the radiation from an ideal black-body surface at the same temperature as given by the Stefan–Boltzmann law. The ratio varies from 0 to 1. The surface of a perfect black-body (with an emissivity of 1—see Kirchoff's definition of a black-body given above) emits thermal radiation at the rate of approximately 448 watts per square metre at 25 °C; all real objects have ε less than 1, and emit radiation at correspondingly lower rates.

The radiant exitance of a surface per unit frequency (ν) or wavelength (λ), $M_{e,\nu}$ and $M_{e,\lambda}$, is defined as watt per square metre per hertz, and watt per square metre per metre, respectively. Dimensionally, we may write: $[M_{e,\nu}] = \mathsf{ML^2T^{-3}/L^2\ T^{-1}} = \mathsf{MT^{-2}}$, or W m^{-2} Hz^{-1} in the SI; $[M_{e,\lambda}] = \mathsf{ML^2T^{-3}/L^2\ L} = \mathsf{ML^{-1}T^{-3}}$, or Wm^{-3} in the SI. Again, there is a difference between the radiant exitance of a perfect black-body and a real body, and this difference is the emissivity. The radiosity (see table 9.2) is the radiant flux leaving (emitted, reflected and transmitted by) a surface per unit area, and has dimensions $\mathsf{MT^{-3}}$, and spectral radiosity is the radiosity of a surface per unit frequency (dimensions, $\mathsf{MT^{-2}}$) or wavelength (dimensions, $\mathsf{ML^{-1}T^{-3}}$), depending on whether the spectrum is taken as a function of frequency, or of wavelength. The radiosity of a surface, J_e, is defined as:

$$J_e = \partial \Phi_e / \partial A = J_{e,em} + J_{e,r} + J_{e,tr}, \tag{10.3}$$

where Φ_e is the radiant flux leaving (emitted, reflected and transmitted) the surface of area A, $J_{e,em} = M_e$ is the emitted component of the radiosity of the surface; that is, its exitance, $J_{e,r}$ is the reflected component of the radiosity of the surface and $J_{e,tr}$ is the transmitted component of the radiosity of the surface. For an opaque surface, $J_{e,tr}$ vanishes, leaving: $J_e = M_e + J_{e,r}$. For a real emitter (not a black-body—sometimes called a grey surface), we may write: $J_e = M_e + J_{e,r} = \varepsilon\sigma T^4 + (1\text{-}\varepsilon)E_e$, where E_e is now the irradiance of the surface.

10.1 Black-body radiation and dimensional analysis

Given that a fourth-power dependence in temperature is something of a rarity, and must have been a surprise back in the late-19th Century, let us see if we can derive something of the physics of what is going on in the Stefan–Boltzmann law by using dimensional analysis. Indeed, this example of dimensional analysis and subsequent examples demonstrate both the power and the weakness of the technique. As mentioned earlier, we first have to consider which variables might be involved in the observation; that is, upon which variables might the Stefan–Boltzmann constant (σ) be dependent. We begin by saying that

$\sigma \propto$ (temperature of black-body)^a (energy content of black-body)^b (the speed of light)^c (the constant of proportionality between radiation and energy)^d.

The dimensions and properties of these variables may be found in table 10.1.

We may rewrite our initial hypothesis of the nature of σ as

$$\sigma \propto (T)^a (k_B)^b (c)^c (h)^d, \tag{10.4}$$

where the variables are defined in table 10.1. Given the dimensions of radiant exitance (see table 9.2), we may write the dimensional equation, which defines the Stefan–Boltzmann constant as:

Table 10.1. Variables for the study of black-body radiation.

Variable	Dimension	Description
σ	MT^{-3}	Stefan–Boltzmann constant, defined as: $\sigma = M_e/T^4$ (Wm^{-2} K^{-4})
T	θ	Temperature of black-body
k_B	$ML^2\,T^{-2}\,\theta^{-1}$	Boltzmann constant
c	LT^{-1}	Speed of light
h	$ML^2\,T^{-1}$	Constant of proportionality between radiation and energy, or the Planck constant
λ	L	Wavelength of radiation
ν	T^{-1}	Frequency of radiation

$$MT^{-3} = (\theta)^a(MT^2T^{-2}\theta^{-1})^b(LT^{-1})^c(ML^2T^{-1})^d = M^{b+d}L^{2b+c+2d}T^{-2--c-d}\theta^{a-b}. \quad (10.5)$$

The solutions are: $a = 4$, $b = 4$, $c =-2$, $d = -3$; and we may write

$$\sigma = K\ T^4k_B^4/c^2h^3, \quad (10.6)$$

where K is the dimensionless constant of proportionality that arises on going from the proportionality of equation (10.4) to the equality of equation (10.5). A comparison of equation (10.5) with the exact result given above demonstrates the limitations of this technique. We have only found the 'bare bones' of the Stefan–Boltzmann constant; we have lost the numerical constants that are dimensionless. In the case of the Stefan–Boltzmann law, this numerical constant is $\frac{2}{15}\pi^5$; that is, a not insignificant factor of about 39. The simplicity and power of the technique of dimensional analysis is clearly revealed in putting the constants in their correct relationship; however, the technique deals only with quantities with dimensions, and so a great deal of fascinating physics is lost.

The Stefan–Boltzmann law has been particularly useful in modelling the temperature of planets. On planets without an atmosphere, the lack of any significant greenhouse effect allows equilibrium temperatures to approach mean surface temperatures; as on Mars, where the equilibrium temperature is 210 K, and the mean surface temperature of emission is 215 K. Consequently, there are large variations in surface temperature over space and time on airless or near-airless bodies like Mars, which has daily surface temperature variations of 50–60 K. Assuming the planet radiates as a black-body (according to the Stefan–Boltzmann law), temperature variations propagate into emission variations to the fourth-power. If the hot object is radiating energy to a cooler surrounding (at a temperature T_c), the net radiated loss is $M_e = \varepsilon\sigma(T^4 - T_c^4)$. This is useful because our understanding of planetary temperatures comes not from direct measurement of the temperatures, but from measurements of energy fluxes. Consequently, in order to derive a meaningful mean surface temperature on an airless body (to compare with an equilibrium temperature), a global average surface emission flux is considered, and then an effective temperature of emission that would produce such a flux is calculated.

The 19th Century research on the measurement of tiny fluctuations of temperature and energy (both of which are related through the constant of Nature known as the Boltzmann constant) led to the development of new technologies for making such measurements. And this advance led, in turn, to the desire to make increasingly precise measurements, which triggered the development of even more precise techniques of measurement, and so on. The scientists' questions were driving the development of technology, by always seeking to push the measurement horizon. This research on temperature showed us how energy flows through matter, and led Max Planck to the concept of the quantum, or the smallest possible quantity of energy, and then to quantum mechanics; arguably the most successful theory ever devised in the physical sciences. Another such pioneer of measurement was Samuel Pierpont Langley (1834–1906); an American astronomer, a founder of aviation, and Secretary of the Smithsonian Institute where he invented the bolometer; an instrument devised to measure the energy of infrared radiation. Today, the bolometer is a well-accepted tool for fundamental investigations in chemical physics and astronomy. However, when the first bolometer was built in Washington City in the late-19th Century, Langley's contemporaries (particularly his friend, the historian Henry Adams) ridiculed him as 'the man attempting to measure nothing'; which was exactly what he was trying to do. Langley was pushing the technology for the measurement of minute changes in temperature beyond its then limits; he was indeed attempting to measure something close to nothing.

For all the joking, Langley made the first measurement of the surface temperature of the Moon by measuring the effect of moonlight (which is principally infrared radiation associated with the glorious vision we see in the visible range of the electromagnetic spectrum) in his bolometer. One can readily imagine the mirth with which his experiment on moonshine must have been greeted by his sophisticated, non-physicist friends. Langley showed that the temperature of the Lunar surface, measured from the intensity of moonlight entering his bolometer varied depending upon the angle of incidence of the moonlight relative to his bolometer; that is, the depth of the Earth's atmosphere that had been traversed by the infrared radiation coming from the Moon (the pathlength of the radiation). Langley went on to show that this angular variation was due to the amount of carbon dioxide in the Earth's atmosphere. In addition, Langley's measurements of the interference of the Sun's infrared radiation incident upon the Earth's surface by carbon dioxide in the Earth's atmosphere was used by the Swedish chemist Svante August Arrhenius (1859–1927) in 1896 to make the first calculation of how the world's climate would change from a future doubling of the levels of carbon dioxide in the atmosphere; scientific research which even our politicians are now realizing is of importance for our future welfare. What seemed laughable and, perhaps, foolish to even the sophisticated and enlightened friends of Langley is today a unique tool for research in the physical sciences. This was a change in perception that occurred in a relatively short period of time. Doubt, questioning and the desire to see what, if anything, is hidden under the noise that is flooding your detector when you make a measurement are far more important than absolute certainty for a scientist. The greater the level of precision with which you examine even a well-known phenomenon often leads to new physics, new discoveries and new insights into the nature of our physical world; and not simply to a more precise value of an already well-known constant.

The observation and subsequent explanation of the Stefan–Boltzmann law led to a widespread investigation of the equilibrium between matter and energy, which was to culminate in the creation of quantum mechanics in 1900. The next stage in this story of the equilibrium between matter and energy concerns the Rayleigh–Jeans law,[1] which is an approximation to the spectral radiance of electromagnetic radiation (see table 9.2), $L_{e,\ \Omega,\ \lambda}$. This is the radiation emitted, as a function of wavelength, by a black-body at a given temperature. Based on classical arguments, the Rayleigh–Jeans law failed but the reasons why it failed led to a new way of looking at Nature.

Ludwig Boltzmann showed that black-body radiation is homogeneous, isotropic (emitted in all directions) and unpolarized, so that $B(\nu,T) = \frac{c}{8\pi}\,\rho(\nu,T)$, where $\rho(\nu,T)$, the spectral density is the energy per unit volume for a frequency ν. Dimensionally, this is: $[B(\nu,T) = M_e] = \mathsf{MT}^{-3} = \mathsf{LT}^{-1}\ (\mathsf{ML}^2\mathsf{T}^{-2}/\mathsf{L}^3) = \mathsf{MT}^{-3}$. Thus, the Stefan–Boltzmann law says (V is the volume of the radiating cavity), $E(T) = V \int \rho(\nu,T)\, d\nu = a\ V\ T^4$. This law was the first thermodynamic consequence derived from the work of Maxwell on electromagnetic theory, according to which the numerical value of the radiation pressure equals one third of the energy per unit volume. This was where the theoretical analysis of radiation stayed until the end of the 19th Century, and the theoretical developments of Max Planck who proposed his radiation law in 1900.

Radiance is the radiant flux emitted, reflected, transmitted or received by a given surface, per unit solid angle per unit projected area (see table 9.2). Spectral radiance is the radiance of a surface per unit frequency, or wavelength depending on whether the spectrum is taken as a function of frequency, or of wavelength. For radiation of wavelength, λ, the Rayleigh–Jeans law may be expressed as

$$L_{e,\Omega,\lambda}(T) = 2ck_BT/\lambda^4, \tag{10.7}$$

where the LHS is the spectral radiance—the power emitted per unit emitting area, per steradian, per unit wavelength; c is the speed of light; k_B is the Boltzmann constant; and T is the absolute temperature. For frequency, ν, the expression becomes

$$L_{e,\Omega,v}(T) = 2v^2k_BT/c^2. \tag{10.8}$$

The Rayleigh–Jeans law agrees with experimental observations at long-wavelengths (low frequencies) but strongly disagrees with observations at short wavelengths (high frequencies). This problem was resolved in 1900 with the derivation by

[1] Named for Lord Rayleigh (the founder of modern dimensional analysis) and Sir James Jeans (1877–1946).

Max Planck of what became known as the Planck law, which gives the correct energy content for radiation at all frequencies.

But let us see how close to the Rayleigh–Jeans law we can get with dimensional analysis. Again, we will assume a number of terms that we believe specify the spectral radiance, L_e (see table 9.2). Given what we know of the Stefan–Boltzmann law, we will assume that the spectral radiance is proportional to the speed of light, the Boltzmann constant, the wavelength of the radiation and the temperature of the emitting black-body (the variables are defined above and in table 9.1):

$$L_{e,\Omega,\lambda} \propto (c)^a (k_B)^b (\lambda)^c (T)^d. \tag{10.9}$$

Thus, we define our dimensional equation,

$$\mathsf{ML^{-1}T^{-3}} = (\mathsf{LT^{-1}})^a (\mathsf{ML^2T^{-2}\theta^{-1}})^b (\mathsf{L})^c (\theta)^d = \mathsf{M}^b \mathsf{L}^{a+2b+c} \mathsf{T}^{-a-2b} \theta^{-b+d}. \tag{10.10}$$

The solutions are: $a = 1$, $b = 1$, $c = -4$ and $d = 1$. Thus, based on our initial assumptions as to the nature of the spectral radiance, we may write our derived form of the wavelength-version of the Rayleigh–Jeans law as

$$L_{e,\Omega,\lambda} = Kck_BT/\lambda^4, \tag{10.11}$$

where K is a constant of proportionality. Our result, via dimensional analysis, is the same as that of Rayleigh and Jeans, except for the numerical factor of 2 (see equation (10.7)). The Rayleigh–Jeans law revealed an important error in the physics of the time. The law predicted an energy output that diverged to infinity as wavelength approaches zero (as frequency tends to infinity). Measurements of the spectral emission of actual black-bodies revealed that the emission agreed with the Rayleigh–Jeans law at low frequencies, but diverged at high frequencies.

The solution to the problem of the equilibrium between matter and energy came in 1900 when Planck published his equation for the energy content of black-body radiation. In terms of wavelength, we may write the Planck law as

$$L_{e,\Omega,\lambda}(T) = \frac{2hc^2}{\lambda^5}\frac{1}{e^{\frac{hc}{\lambda k_B T}} - 1} \tag{10.12}$$

where h is the Planck constant and k_B the Boltzmann constant. The Planck law does not diverge at short wavelengths, and agrees well with the experimental data, but its full significance (which ultimately led to quantum theory) was only appreciated several years later. As we can expand the exponential term in equation (10.12), we can write

$$\exp(hc/\lambda k_B T) - 1 = 1 + \left(\frac{hc}{\lambda k_B T}\right) + \ldots - 1 \tag{10.13}$$

and so

$$\frac{1}{\exp(hc/\lambda k_B T) - 1} = \frac{\lambda k_B T}{hc} \tag{10.14}$$

This results in Planck's radiation law reducing to

$$L_{e,\Omega,\lambda}(T) = (2ck_BT/\lambda^4), \tag{10.15}$$

which is identical to the classically derived Rayleigh–Jeans expression. The same argument can be applied to the black-body radiation expressed in terms of frequency $\nu = c/\lambda$. In the limit of small frequencies; that is, $h\nu << k_BT$,

$$L_{e,\Omega,\lambda}(T) = \frac{2h\nu^3}{c^2}\frac{1}{e^{\frac{h\nu}{\lambda k_BT}}-1} \approx \frac{2h\nu^3}{c^2}\cdot\frac{k_BT}{h\nu} = \frac{2\nu^2k_BT}{c^2} \tag{10.16}$$

This is the Rayleigh–Jeans law in the limit of small frequencies, as seen in equation (10.8).

As the use of dimensional analysis to explore the well-known physics of black-bodies has been revealing, let us consider a fuller use of dimensional analysis. This time we will consider the intensity, I, of emitted radiation is a function of:

$$I = (\lambda)^a(k_B)^b(h)^c(c)^d(T)^e; \tag{10.17}$$

where I is the spectral radiance, $L_{e,\Omega,\lambda}$ (see table 9.2), with dimensions $\mathsf{ML^{-1}T^{-3}}$. The other variables are as defined in table 10.1. There are six variables containing four base dimensions; thus, by the Buckingham Π-theorem there are two nondimensional groups to be identified. Thus, $\Pi = (I)\,(\lambda)\,(k_B)\,(h)\,(c)\,(T)$, where we will use (λ) (k_B) (h) (c) as the repeating variables. Thus,

$$\Pi_1 = (I)\,(\lambda)^a(k_B)^b(h)^c(c)^d = \mathsf{ML^{-1}T^{-3}(L)^a(ML^2T^{-2}\theta^{-1})^b(ML^2T^{-1})^c(LT^{-1})^d}. \tag{10.18}$$

This equation generates four linear equations, one for each base dimension.

$$\mathsf{M}: b + c + 1 = 0$$

$$\mathsf{L}: a + 2b + 2c + d - 1 = 0$$

$$\mathsf{T}: -2b - c - d - 3 = 0$$

$$\theta: -b = 0.$$

Immediately, we see that $b = 0$. The other exponents are: $a = 5$, $c = -1$ and $d = -2$. Giving $\Pi_1 = I\lambda^5/hc^2$. Using the repeat variables to find Π_2, which involves θ, we have:

$$\Pi_2 = \theta(\lambda)^a(k_B)^b(h)^c(c)^d = \theta\mathsf{(L)^a(ML^2T^{-2}\theta^{-1})^b(ML^2T^{-1})^c(LT^{-1})^d}. \tag{10.19}$$

This equation generates four linear equations, one for each base dimension:

$$\mathsf{M}: b + c = 0$$

$$\mathsf{L}: a + 2b + 2c + d = 0$$

$$\mathsf{T}: -2b - c - d = 0$$

$$\theta: -b + 1 = 0.$$

This time, the solutions are: $a = 1$, $b = 1$, $c = -1$ and $d = -1$. Giving $\Pi_2 = T\lambda k_B/ h.c.$ Within the Buckingham Π-theorem we may write $\Pi_1 = f(\Pi_2)$, which gives:

$$\Pi_1 = I\lambda^5/hc^2 = f(T\lambda k_B/h.\ c.\ ),$$

$$I = T\lambda k_B hc^2/\lambda^5 hc = cTk_B/\lambda^4. \tag{10.20}$$

This result agrees with the full expression seen in equation (10.15), except for the numerical factor.

10.2 The displacement law of Wilhelm Wien

The displacement law of German physicist, Wilhelm Wien (1864–1928), states that the black-body radiation curve for different temperatures peaks at a wavelength inversely proportional to the temperature (see figure 10.2) [3]. The shift of that peak is a direct consequence of the Planck radiation law which describes the spectral radiance of black-body radiation as a function of wavelength ($L_{e,\,\Omega,\,\lambda}$) at any given temperature. However, it had been discovered by Wilhelm Wien several years before Max Planck developed the more general equation, his radiation law, which describes the entire shift of the spectrum of black-body radiation toward shorter wavelengths as temperature increases.

Wien's displacement law can be derived from Planck's law. The wavelength and the frequency forms of Planck's law are, respectively:

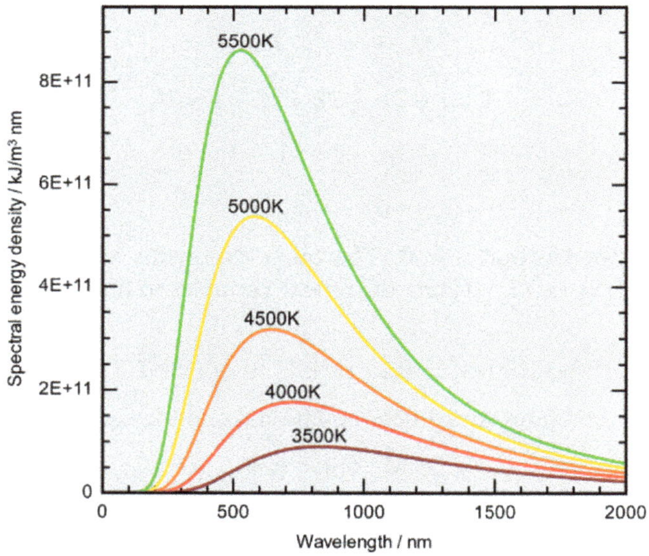

Figure 10.2. Black-body radiation as a function of wavelength for various temperatures. Each temperature curve peaks at a different wavelength and Wien's law describes the shift of that peak. This 'Wiens law' image has been obtained by the author from the Wikimedia website where it was made available under a CC BY-SA 3.0 license. It is included within this article on that basis. It is attributed to 4C.

$$\rho(\lambda, T) = \frac{2hc^2}{\lambda^5}\frac{1}{e^{\frac{hc}{\lambda k_B T}} - 1}, \text{ and}$$
$$\rho(\nu, T) = \frac{2h\nu^3}{c^3}\frac{1}{e^{\frac{h\nu}{k_B T}} - 1}, \tag{10.21}$$

where $\rho(\lambda,T)$ and $\rho(\nu,T)$ define energy density (energy per unit volume per unit wavelength and per unit frequency, respectively). Differentiating the expression for the frequency form,

$$\frac{d}{d\nu}\rho(\nu, T) = \frac{d}{d\nu}\left(\frac{2h\nu^3}{c^3}\frac{1}{e^{\frac{h\nu}{k_B T}} - 1}\right) = 0$$
$$= \frac{2h}{c^3}\frac{d}{d\nu}\left(\frac{\nu^3}{e^{\frac{h\nu}{k_B T}} - 1}\right) = 0$$

Apply the product rule (the power rule and the chain rule), and make the substitution, $u = h\nu/k_B T$; we eventually find $\nu \approx 2.8214\ k_B T/h \approx (5.8 \times 10^{10}\ s^{-1}\ K^{-1})T$. The consequence is that the shape of the black-body radiation function shifts proportionally in frequency with temperature. When Max Planck later formulated the correct black-body radiation function it did not include Wien's constant explicitly. Rather, Planck's constant h was created and introduced into his new formula. But Wien's constant ($u = h\nu/k_B T$) can be obtained from h and k_B.

Following on from the concept of dimensional homogeneity, one can look at the dimensions of energy density, and see how two quantities from different areas of physics can have the same dimensions. Imagine a volume of space of cross-sectional area A ($[A] = \mathsf{L}^2$). This area is being traversed by a fluid moving at a velocity v ($[\mathrm{v}] = \mathsf{LT}^{-1}$). This fluid could be light, in which case v is the speed of light, c, or it could be a liquid, in which case v can take a wide-range of values depending on the driving force for the flow and the geometry. For both light and a fluid, the energy density of the volume created by considering the amount of light or fluid moving across the area in unit time (the flux of the fluid is proportional to $A\ c$ for light, and $A\ v$ for a fluid) is energy/area × velocity = $\mathsf{ML^2T^{-2}/L^2\ LT^{-1}} = \mathsf{ML^{-1}T^{-1}}$.

In chapter 5, we considered a moving fluid with the velocity field varying in one dimension; for example, y, this can be expressed as $\tau \propto \frac{dv}{dy}$. Here, v is the fluid velocity, τ is the shear stress in the fluid (which has the same dimensions as pressure; that is, force/area), and the derivative is of the velocity component parallel to the direction of shear, relative to displacement in the perpendicular direction. The constant of proportionality is the dynamic viscosity, μ; thus, $\tau = \mu\ \frac{dv}{dy}$. The dimensions of this property are: $[\tau] = \mathsf{MLT^{-2}/L^2} = \mu\ \frac{dv}{dy} = \mu\ (\mathsf{LT^{-1}.\ L^{-1}})$; and so $[\mu] = \mathsf{MLT^{-2}.\ L^{-2}/\ LT^{-1}.\ L^{-1}} = \mathsf{ML^{-1}T^{-1}}$ (kg $m^{-1}s^{-1}$ in the SI).

For black-body or cavity radiation the frequency density is $\rho(\nu)$, which also has dimensions $\mathsf{ML^{-1}T^{-1}}$. If we let $\rho(\nu) = (h)^a\ (\nu)^b\ (c)^c$; where h is the Planck constant, ν is the frequency of the radiation being considered, and c the speed of light, we may write the dimensional equation as $[\rho(\nu)] = \mathsf{ML}^{-1}\mathsf{T}^{-1} = (\mathsf{ML}^2\mathsf{T}^{-1})^a(\mathsf{T}^{-1})^b(\mathsf{LT}^{-1})^c = \mathsf{M}^a\ \mathsf{L}^{2a+c}\mathsf{T}^{-a-b-c}$. The solutions are: $a = 1$, $b = 3$ and $c = -3$; that is, $[\rho(\nu)] = \mathsf{ML}^{-1}\mathsf{T}^{-1} = h\nu^3/c^3 = (h\nu)\ \nu^2/c^3 = (h\nu)/cL^2$ (where L is a length variable, because $[\nu/c] = \mathsf{L}^{-1}$). Alternatively, we could look at this analysis and point out that if you count the modes per unit frequency per unit volume, the answer is $8\pi\nu^2/c^3$, which when you multiply by an energy, $\mathsf{ML}^2\mathsf{T}^{-2}$ (in this case, $h\nu$), we obtain $\mathsf{ML}^{-1}\mathsf{T}^{-1}$. For the fluid mechanical example, the energy multiplying the dynamic contents of the volume would be $\frac{1}{2}mv^2$. In both cases, we are considering the distribution of energy in a volume per unit time; but we generate very different physics by considering the origin of the distributed energy. Indeed, it was by analyzing the distribution of energy in a volume of space that Max Planck moved from classical physics to quantum physics, and formulated the idea of the quantum.

The radiation law first derived by Rayleigh used principles of classical statistics, with a correction for a numerical factor by Jeans. It fulfilled both Kirchhoff's findings [2] and the requirements of Wien's conventional displacement law, $\rho(\nu, T) \propto \nu^3 f(\nu, T)$, or $\rho(\lambda, T) \propto \lambda^5 f(\lambda, T)$. It is called the classical black-body radiation law because it is restricted to the two constants c and k_B which are well-known in classical physics. When the frequency is small and the temperature relatively high, this formula works well (indeed it can be considered as an asymptotic solution for $\nu \to 0$). But it is obvious that the Rayleigh–Jeans law cannot be correct, because for $\nu \to \infty$ the monochromatic energy density, $\rho(\nu, T)$, would tend to infinity. In 1911, Ehrenfest [4] called this divergence, the Rayleigh–Jeans ultraviolet catastrophe. Interestingly, Edgar Buckingham published a derivation of the Wien displacement law via a dimensional analysis in 1912 [5]. Evidently, dimensional analysis was being used routinely to check the theories being developed.

10.3 The cosmic microwave background

The Cosmic Microwave Background was first predicted in 1948 by Ralph Alpher and Robert Herman, who estimated the temperature of the cosmic microwave background to be 5 K. Although there were several previous estimates of the temperature of space, these suffered from two major shortcomings. First, they were measurements of the effective temperature of space, and did not suggest that space was filled with a thermal Planck spectrum of radiation. Secondly, the previous estimates depended on our being at a special spot at the edge of the Milky Way galaxy, and they did not suggest the distribution of the radiation to be isotropic. The estimates yielded very different predictions, depending upon where the Earth happened to be inside the universal black-body.

The interpretation of the cosmic microwave background was a controversial issue in the 1960s, with some proponents of the steady-state theory arguing that the microwave background was the result of scattered starlight. However, during the 1970s, the consensus became that the cosmic microwave background is a remnant of the Big Bang. This was largely because new measurements, at a range of frequencies, showed that the spectrum was a perfect thermal, black-body spectrum—a result that the steady-state model was unable to reproduce. The cosmic microwave background

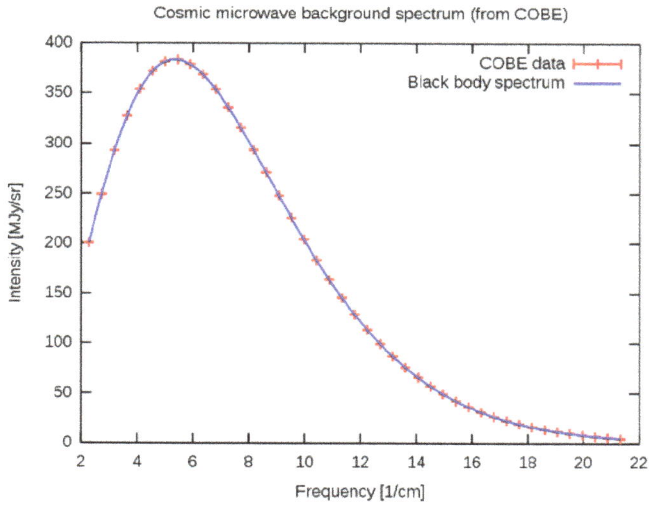

Figure 10.3. Graph of cosmic microwave background spectrum measured by spectrometers aboard the Cosmic Background Explorer satellite, the most precisely measured black-body spectrum. The error bars are too small to be seen even in an enlarged image, and it is impossible to distinguish the observed data from the theoretical curve. Reproduced from https://en.wikipedia.org/wiki/Cosmic_microwave_background. Image stated to be in the public domain. It is included within this article on that basis.

radiation is an emission of uniform, black-body thermal energy coming from all parts of the sky, see figure 10.3. The radiation is isotropic to about one part in 10^5.

In the Big Bang model for the formation of the Universe, inflationary cosmology predicts that after about 10^{-37} s the nascent universe underwent exponential growth that smoothed out nearly all irregularities. The remaining irregularities were caused by quantum fluctuations in the inflation field that caused the inflation event. Long before the formation of stars and planets, the early-universe was smaller, much hotter and, starting 10^{-6} s after the Big Bang, filled with a uniform, interacting plasma of photons, electrons and baryons. This primordial radiation field is a prediction from a Big Bang universe—that in its earliest stages, the Universe was at a high enough temperature to be fully ionised when processes such as Thompson scattering and Bremsstrahlung efficiently thermalised the radiation field. Assuming an adiabatic expansion of the Universe, one would then expect to observe a radiation field, which would have retained the black-body spectrum, but at a much lower temperature.

As the Universe expanded, adiabatic cooling caused the energy density of the plasma to decrease until it became possible for electrons to combine with protons, forming hydrogen atoms. This recombination event happened when the temperature was around 3000 K or when the Universe was approximately 379 000 years old. As photons did not interact strongly with electrically-neutral atoms, the former began to travel freely through space, resulting in the decoupling of matter and radiation. The colour temperature, that is, the temperature of an ideal black-body radiator that radiates light of a colour comparable to that of the light source of the ensemble of decoupled photons, has continued to fall ever since. It is presently at 2.7260 K ±0.0013 K, and it will continue to drop as the Universe expands. According to the Big Bang model, the radiation from the sky we measure today comes from a spherical surface called the 'surface of last scattering.' This represents the set of locations in space at which the decoupling event is estimated to have occurred and at a point in time such that the photons from that distance have just reached observers. Most of the radiation energy in the Universe is in the cosmic microwave background, making up a fraction of roughly 6×10^{-5} of the total density of the Universe. Two of the greatest successes of the Big Bang theory are its prediction of the almost perfect black-body spectrum (see the agreement between observation and theory in figure 10.3), and its detailed prediction of the small, but observable anisotropies in the cosmic microwave background [6]. We exist within an expanding black-body.

References

[1] Kirchhoff G 1860 Ueber das Verhältniss zwischen dem Emissionsvermögen und dem Absorptionsvermögen der Körper für Wärme and Licht *Ann. Phys. Chem.* **109** 275–301
Kirchhoff G 1860 On the relation between the radiating and absorbing powers of different bodies for light and heat *Philos. Mag. J. Sci* Series 4 **20** 1–21

[2] Williams J H 2020 *Defining and Measuring Nature: The Make of All Things* 2nd edn (Bristol: IOP Publishing)

[3] Wien W 1894 Temperatur und Entropie der Strahlung (On the temperature and entropy of radiation) *Ann. Phys.* **52** 132–65

[4] Ehrenfest P 1911 Welche Züge der Lichtquantenhypotheses spielen in der Theorie der Wärmestrahlung eine wesentliche Rolle? (Which features of the light quantum hypotheses play an essential role in the theory of heat radiation?) *Ann. Phys.* **36** 91–118

[5] Buckingham E 1912 On the deduction of the Wien displacement law *Bulletin of the Bureau of Standards* **8** 545–57

[6] White M 1999 Anisotropies within the CMB *Proc. of the Los Angeles Meeting* DPF 99, UCLA. arXiv: astro-ph/9903232

IOP Publishing

Dimensional Analysis
The great principle of similitude
Jeffrey H Williams

Chapter 11

Dimensions involving molecules and fields

If it is love that makes the world go round; it is self-induction that makes electromagnetic waves go round the world.

Oliver Heaviside (1850–1925)

In chapter 9 we looked at molecular polarizabilities; that is, the general response of matter to an applied electric field (at optical frequencies, or a static field). In such experiments, the applied field (E) is generally weak, and the response of the material is therefore linear in the applied field ($Response \propto E$). If, however, the applied field is intense, then the response of the matter may become non-linear, and there are hyperpolarizabilities that describe the electric dipole induced by E^2 and E^3. And we saw how these hyperpolarizabilities are responsible for collision-induced, or pressure-induced light scattering in liquids. Below, we will see how they also provide an example of dimensional homogeneity.

11.1 Polarization and magnetization

But what happens when we apply an electric field to a condensed phase? We will assume we remain in the weak-field limit; that is, a linear response to the applied field. Under these conditions, the applied field will polarize the medium by orienting the charges and dipole moments already present. Even though these internal charge separations are small in magnitude, their number and proximity will modulate the influence of the applied field—the electric field atomically-close to a molecular dipole moment is larger than any applied laboratory electric field. Consequently, the resultant polarization, $\boldsymbol{P}$, of the solid will add a contribution to the electric field, $\boldsymbol{E}$, that is being applied to the solid. We are assuming a linear relationship: $\boldsymbol{P} = \chi_{\mathrm{E}}(\boldsymbol{E})\,\boldsymbol{E}$, where $\chi_{\mathrm{E}}(\boldsymbol{E})$ is the electric susceptibility—analogous to the polarizability of dispersed matter. However, we cannot simply add $\boldsymbol{E}$ and $\boldsymbol{P}$, because they do not have the same dimensions. The polarization has units of Cm m^{-3} = Cm^{-2} = IT L^{-2}, which means the electric susceptibility must have units of C^2 Nm^{-2} ($M^{-1}L^{-3}T^4I^2$), which are the dimensions of ε_0, the vacuum permittivity. Consequently, to obtain the effect of the polarization on the applied field we multiply $\boldsymbol{E}$ by ε_0 and add the result to $\boldsymbol{P}$ to give $\boldsymbol{D}$, the electric displacement,

$$\boldsymbol{D} = \varepsilon_0 \boldsymbol{E} + \boldsymbol{P} \tag{11.1}$$

$$= (\varepsilon_0 + \chi_E)\boldsymbol{E} = \varepsilon \boldsymbol{E}, \tag{11.2}$$

where ε is the permittivity of the medium (a property determined by the chemical composition and, in a solid by the crystal structure). With this definition, $\chi_E \rightarrow 0$ and $\varepsilon \rightarrow \varepsilon_0$ in vacuum. It is important to recognize that $\boldsymbol{E}$ is the fundamental field—it generates $\boldsymbol{P}$. If we write $\boldsymbol{E}$ as $\boldsymbol{E} = (1/\varepsilon_0)(\boldsymbol{D} - \boldsymbol{P})$, we may relate $\boldsymbol{D}$ with free charges, and the polarization, $\boldsymbol{P}$, is related to the bound charges in the medium, which are associated with localised charges (e.g. dipole moments). The subject of dielectric polarization is vast, and there are many textbooks explaining the phenomena; for example, [1]. For media that responds linearly to applied fields, we usually do not use χ_E, but a dimensionless quantity, κ, the dielectric constant, which is related to the permittivity, $\varepsilon = \varepsilon_0 \kappa$. This leads to $\boldsymbol{D} = \kappa \varepsilon_0 \boldsymbol{E} = \varepsilon \boldsymbol{E}$; where, for most media, $\kappa > 1$.[1] These and other quantities we will come to below are summarized in table 11.1; and their interconnectivity is shown in figure 11.1.

Magnetic interactions are generally more complicated than electric analogues (as the old saying has it: Electric fields are generated by static charges, while magnetic fields are generated by moving charges). There are three distinct regimens of magnetic interactions: ferromagnetism, paramagnetism and diamagnetism, in order of decreasing strength. Ferromagnetism and paramagnetism lead to a magnetization of the material upon application of an external magnetic field as a result of the alignment of permanent magnetic dipoles. Diamagnetism, on the other hand, is similar to polarization in dielectric materials in that a nonpermanent magnetization is induced in the medium, and when the magnetic field is removed, the induced magnetization disappears. Paramagnetism and diamagnetism can be treated within the weak field limit, as with polarization; that is, the induced magnetization, $\boldsymbol{M}$, is (where χ_M is dimensionless—see table 11.1) $\boldsymbol{M} = \chi_M \boldsymbol{H}$; where χ_M is the magnetic susceptibility and $\boldsymbol{H}$ is the applied magnetic field. However, in ferromagnetic materials (permanent magnets), $\boldsymbol{M}$ is a strong function of $\boldsymbol{H}$.

As in dielectrics, we also have two fields in the case of magnetic interactions. The term 'magnetic field' is confusingly used for two distinct, but closely-related vector fields denoted by the symbols $\boldsymbol{B}$ and $\boldsymbol{H}$ (see table 11.1). In the SI, $\boldsymbol{H}$ (magnetic field strength, or magnetic field intensity), has units of ampere per metre (A m^{-1}). $\boldsymbol{B}$ (magnetic flux density, or magnetic induction) is measured in tesla, or kg s^{-2} A^{-1}, which is equivalent to newton per metre per ampere (see table 11.1). In the c.g.s.

[1] Water, which has a large permanent electric dipole moment, undergoes significant polarization in the presence of an applied field, and has a large dielectric constant of 78.4. Benzene, on the other hand, has a dielectric constant of 2.3. The ceramic ferroelectric, barium titanate, has a dielectric constant of order 10^5 at elevated temperatures.

Table 11.1. The six electromagnetic fields, and the permittivity and permeability of space (see figure 11.1 for details of how these quantities are related).

Name (symbol)	SI unit	Dimension
Electric field (***E***)	$\mathrm{Vm^{-1} = N\ C^{-1}}$	$\mathrm{M\ L\ T^{-3}\ I^{-1}}$
Polarization (***P***): The density of permanent or induced electric dipole moments in a dielectric material	$\mathrm{Cm^{-2}}$	$\mathrm{I\ T\ L^{-2}}$
Electric displacement (***D***)	$\mathrm{Cm^{-2}}$	$\mathrm{I\ T\ L^{-2}}$
Magnetic induction (***B***)	$\mathrm{N\ A^{-1}\ m^{-1}}$	$\mathrm{M\ T^{-2}\ I^{-1}}$
Magnetization (***M***): The magnetic dipole moment per unit volume	$\mathrm{A\ m^{-1}}$	$\mathrm{I\ L^{-1}}$
Magnetic field intensity (***H***)	$\mathrm{A\ m^{-1}}$	$\mathrm{I\ L^{-1}}$
Permittivity	ε_0	$\mathrm{M^{-1}L^{-3}T^4I^2}$
Permeability	μ_0	$\mathrm{MLT^{-2}\ I^{-2}}$

Electric field (E) $\mathrm{V\ m^{-1}} =$ $\mathrm{kg\ m\ s^{-3}\ A^{-1}}$ $\mathrm{(MLT^{-3}I^{-1})}$	Multiplied by	ε_0 $\mathrm{kg^{-1}m^{-3}\ s^4}$ $\mathrm{A^2}$ $\mathrm{(M^{-1}L^{-3}T^4I^2)}$	equals	**Electric flux density (D)** $\mathrm{m^{-2}\ s\ A}$ $\mathrm{(L^{-2}\ I\ T)}$
Equals				Multiplied by
Velocity of light (*c*) $\mathrm{m\ s^{-1}}$ $\mathrm{(LT^{-1})}$				Velocity of light (*c*) $\mathrm{m\ s^{-1}}$ $\mathrm{(LT^{-1})}$
Multiplied by				Equals
Magnetic flux density (B) $\mathrm{kg\ s^{-2}\ A^{-1}}$ $\mathrm{(MT^{-2}\ I^{-1})}$	equals	μ_0 $\mathrm{kg\ m\ s^{-2}\ A^{-2}}$ $\mathrm{(MLT^{-2}\ I^{-2})}$	Multiplied by	**Magnetic field (H)** $\mathrm{m^{-1}\ A}$ $\mathrm{(L^{-1}\ I)}$

Figure 11.1. The central elements in the dimensions of electromagnetic quantities.

system, the unit of the $\boldsymbol{H}$ field is the oersted and the unit of the $\boldsymbol{B}$ field is the gauss. In the SI, the unit ampere per metre ($\mathrm{A\ m^{-1}}$), which is equivalent to newton/weber, is used for the $\boldsymbol{H}$ field and the unit of tesla is used for the $\boldsymbol{B}$ field. The oersted is defined as a dyne per unit pole. The oersted is $1000/4\pi$ ($\approx$79.577 4715) amperes per metre, in terms of SI units. The $\boldsymbol{H}$ field strength inside a long solenoid wound with 79.58 turns per metre of a wire carrying 1 A is approximately 1 oersted. The preceding statement is exactly correct if the solenoid considered is infinite in length with the current evenly distributed over its surface. The oersted is closely related to the gauss, the c.g.s. unit of magnetic flux density, $\boldsymbol{B}$. In a vacuum, if the magnetizing field strength is 1 Oe, then the magnetic field density is 1 G, whereas, in a medium having permeability μ_r (relative to permeability of vacuum), their relation is: $\boldsymbol{B}(\mathrm{G}) = \mu_r\ \boldsymbol{H}(\mathrm{Oe})$.

As with $\boldsymbol{E}$ and $\boldsymbol{D}$, the $\boldsymbol{H}$ and $\boldsymbol{B}$ fields differ in their relation with matter—magnetization. In a vacuum, the two fields are related through the vacuum permeability, μ_0; that is, $\boldsymbol{B}/\mu_0 = \boldsymbol{H}$; but in a magnetized material, the terms differ by the material's magnetization at each point. By analogy with the above analysis for dielectrics, $\boldsymbol{H}$ may be written in terms of $\boldsymbol{M}$, the magnetization of the medium, and the permeability of the vacuum, μ_0,

$$\boldsymbol{H} = (1/\mu_0)\boldsymbol{B} - \boldsymbol{M}. \tag{11.3}$$

From table 11.1 we see that $\boldsymbol{B}$ has dimensions $\mathsf{MT^{-2}\ I^{-1}}$, while $\boldsymbol{M}$ has dimensions of $\mathsf{IL^{-1}}$. Since μ_0 has dimensions of $\mathrm{NA^{-2}} = \mathsf{MLT^{-2}\ I^{-2}}$, $\boldsymbol{M}$ and $\boldsymbol{B}/\mu_0$ have the same dimensions. We can then write

$$\boldsymbol{B} = \mu_0(\boldsymbol{H} + \chi_{\mathrm{M}}\boldsymbol{H}) = \mu_0(1 + \chi_{\mathrm{M}})\boldsymbol{H}, \tag{11.4}$$

and as in dielectrics, when we are dealing with a particular substance, $\mu = \mu_0(1 + \chi_{\mathrm{M}})$.[2] Paramagnetic materials have positive magnetic susceptibility, and diamagnetic materials have negative magnetic susceptibility. Thus, $\boldsymbol{M}$ points in the opposite direction from the applied field in the two cases.

Unlike dielectrics, magnetic susceptibility, or volume magnetic susceptibility, is a dimensionless proportionality constant that indicates the degree of magnetization of a material in response to an applied magnetic field ($\boldsymbol{M} = \chi_{\mathrm{v}}\ \boldsymbol{H}$). There are two other widely-used measures of susceptibility: the mass magnetic susceptibility (χ_{mass} or χ_{m}), measured in $\mathrm{m^3\ kg^{-1}}$ (in the SI), and the molar magnetic susceptibility (χ_{mol}) measured in $\mathrm{m^3\ mol^{-1}}$ that are defined below, where ρ is the density in $\mathrm{kg\ m^{-3}}$ and M is molar mass in $\mathrm{kg\ mol^{-1}}$: $\chi_{\mathrm{m}} = \chi_{\mathrm{v}}/\rho$ and $\chi_{\mathrm{mol}} = M\ \chi_{\mathrm{mass}} = M\ \chi_{\mathrm{v}}/\rho$. These definitions are according to SI conventions. In the c.g.s. system, these quantities rely on a different definition of the permeability of free space:

$$\boldsymbol{B}^{\mathrm{cgs}} = \boldsymbol{H}^{\mathrm{cgs}} + 4\pi\boldsymbol{M}^{\mathrm{cgs}} = (1 + 4\pi\chi_{v^{\mathrm{cgs}}})\boldsymbol{H}^{\mathrm{cgs}}. \tag{11.5}$$

The dimensionless c.g.s. value of volume susceptibility is multiplied by 4π to give the dimensionless SI volume susceptibility: $\chi_{\mathrm{v}}^{\mathrm{SI}} = 4\pi\ \chi_{\mathrm{v}}^{\mathrm{cgs}}$. Values of c.g.s. mass

[2] The definition of the permeability has recently changed, due to the creation of the Quantum-SI. Table III contains details of how the new definition of the quantity of electricity has changed the definition of μ_0 [2].

susceptibilities are usually given in cm^3 $gram^{-1}$ or emu g^{-1} Oe^{-1}; so, to convert to the SI volume susceptibility we write: $\chi_v{}^{SI} = 4\pi\, \rho^{cgs}\, \chi_m{}^{cgs}$, where ρ^{cgs} is the c.g.s. density (g cm^{-3}), so: $\chi_v{}^{SI} = 4\pi\, 10^{-3}\, \rho^{SI}\, \chi_m{}^{cgs}$. The molar susceptibility is measured in cm^3 mol^{-1} or emu mol^{-1} Oe^{-1} in c.g.s., and is converted by considering the molar mass (the mass of a sample of that compound divided by the amount of substance in that sample, measured in moles).

The ***B*** field can also be defined by the torque on a magnetic dipole, ***m***. If a magnet is placed in an external ***B*** field, it will experience a torque (as would an electric dipole placed in an electric field)[3]. This torque is the rotational equivalent of linear force; that is, N m or kg m^2 s^{-2}, with dimensions ML^2T^{-2}. The magnitude of the torque depends on the orientation of the magnet with respect to the magnetic field. There are two oppositely-directed orientations in which the magnet will experience the greatest torque, and the magnitude of the magnetic moment (***m***) is defined as the maximum torque experienced by the magnet when placed in unit external field. The magnitude and direction of the torque, $\boldsymbol{\tau}$, is given by, $\boldsymbol{\tau} = \boldsymbol{m} \times \boldsymbol{B}$. The SI unit for magnetic moment is N m T^{-1} (where T is the SI derived unit for magnetic flux density, the tesla, see figure 11.1; in Gaussian-c.g.s. units, ***B*** is measured in gauss (G), with a conversion factor: 1 T = 10^4 G). If an electric current *I* flows in a flat coil of area ***A*** (recall that area is a vector quantity), the torque experienced in a magnetic field is given by $\boldsymbol{\tau} = \boldsymbol{I}\, \boldsymbol{A} \times \boldsymbol{B}$. This means that the magnetic moment of the coil is given by $\boldsymbol{m} = I\boldsymbol{A}$. Thus, the unit A m^2 (IL^2) is also a correct SI unit for magnetic moment, though the concept of current in a coil needs to be emphasized in a particular context; in fact, A m^2 = Nm T^{-1} = J T^{-1}. While J T^{-1} is also formally dimensionally correct, and is to be found in many sources, it is perhaps better to restrict the unit joule to work or energy, and to use N m for torque. Although these are dimensionally similar, they are conceptually different.

When we consider the influence of an electric field on dispersed matter, we say that the polarizability (α) is the response of the matter to the electric field (***E***); this response is observed as an induced electric dipole moment ($\boldsymbol{\mu}$). The magnetic field (***H***), acts analogously on dispersed matter to generate a magnetic dipole (***m***) via the magnetic susceptibility, χ (but the effect is much weaker, by order, *c*, the speed if light). Thus, $\mu = \alpha E$ and $m = \chi H$. The magnetic susceptibility, like the polarizability, is a property of all matter. Thus, argon atoms have a (dia)magnetic susceptibility and an electric polarizability; $\chi_{Ar} = -19.6 \times 10^{-6}$ cm^3 $mole^{-1}$ and $\alpha_{Ar} = 1.664 \times 10^{-24}$ cm^3, respectively. These atomic properties are given in the c.g.s. system of units and can be seen to be volumes. As mentioned above, the c.g.s. unit of polarizability is a volume, and the magnitude of the polarizability scales with the volume of the molecule and the number of electrons.

Thus, $[\alpha]_{cgs} = L^3$, and to convert to the SI we multiply by the permittivity of free space, ε_0. We write, $[\alpha]_{SI} = [\alpha]_{cgs}\, 4\pi\varepsilon_0 = L^3\,(M^{-1}\, L^{-3}\, T^4\, I^2) = M^{-1}T^4I^2$, or C^2 m^2 J^{-1} in the SI. With the magnetizability, we would have: $[\chi]_{cgs} = \frac{[m]}{[H]} = L^2I/L^{-1}I = L^3$. And to

[3] The torque, τ, exerted by a ***B*** field on a magnetic dipole is equivalent to the torque exerted by an electric field on an electric dipole. Dimensionally: (for the magnetic case), $[\tau]$ = [magnetic dipole] [***B*** field] = IL^2 $MT^{-2}I^{-1} = ML^2T^{-2}$; (for the electric case), $[\tau]$ = [electric dipole] [electric field] = LIT $MLT^{-3}I^{-1} = ML^2T^{-2}$.

convert to the SI, we need to divide by the permeability of the vacuum: $[\chi]_{SI} = (1/\mu_0)$ $[\chi]_{cgs} = L^3/MLT^{-2}\ I^{-2} = M^{-1}L^2T^2I^2$. The SI unit for the magnetic susceptibility is JT^{-2} (joule tesla^{-2}); that is, $JT^{-2} = \frac{[\text{Energy}]}{[\text{Tesla}][\text{Tesla}]} = ML^2T^{-2}/(M\ T^{-2}\ I^{-1})^2 = M^{-1}\ L^2T^2I^2$. The SI unit of diamagnetic magnetic susceptibility, joule tesla^{-2}, may be converted to the c.g.s. unit of erg gauss^{-2}; $10 \times 10^{-29}\ JT^{-2} = 10^{-30}$ erg gauss^{-2}.

11.2 Electromagnetic fields

A static charge distribution, as represented by a molecule, has a potential energy represented by the work that would have been done against the electrostatic forces of interaction in building up the distribution from the state in which the constituent charges are infinitely dispersed. There is a close analogy with the potential energy of a mass distribution; the difference being that the gravitational energy is always > 0 since the gravitational force is always attractive. The gravitational force varies with distance as r^{-2}, so a change in coordinate system does not change the phase of the interaction. The electrostatic force, on the other hand, can be positive or negative, and can cancel itself out.

Consider two point-charges, e_1 and e_2; originally infinitely far apart, they have been brought together to a separation of r. The work done against the Coulomb force is

$$e_1e_2/(4\pi\varepsilon_0)r. \tag{11.6}$$

If e_1 is held fixed, the work done is e_2 times the potential of e_1. Thus, equation (11.6) is the potential energy. We may write that the work done in setting up a static charge distribution (*in vacuo*) from infinitely dispersed charges is:

$$½\int \rho(\varphi)d\tau = ½\varepsilon_0 \int \mathbf{E}^2 d\tau. \tag{11.7}$$

The RHS of equation (11.7) is the field energy density; energy divided by volume; that is J m^{-3} (ML^2T^{-2}/L^3). Dimensionally, we may write, $[\varepsilon_0] = M^{-1}L^{-3}\ T^4\ I^2$, and $[E] = ML\ \ T^{-3}\ I^{-2}$; and the dimensions of the field energy density are $[\varepsilon_0]$ $[E]^2 = ML^{-1}T^{-2}$. There will also be a smaller (by a factor of c, where c is the speed of light) energy density for the vacuum magnetic field, equivalent to a steady current $½\mu_0\boldsymbol{H}^2$. It is the combination of the large electric field energy density, and the smaller magnetic field energy density that gives the Poynting vector, $\mathbf{S} = \mathbf{E} \times \mathbf{H}$. This product of the two energy densities specifies the power flux density intrinsic to the electromagnetic field.

For a plane wave, the intensity, I, (power per unit area; see table 11.1) has a magnitude: $I = \boldsymbol{E}\ \boldsymbol{B}/\mu_0$, where $B = E/c$, and so the magnitude of the instantaneous intensity is:

$$I = E^2/c\mu_0 = cB^2/\mu_0. \tag{11.8}$$

The average intensity is:

$$I_{av} = E_{rms}B_{rms}/\mu_0 = E_{max}B_{max}/2\mu_0 = E_{max^2}/2c\ \ \mu_0 = c\ \ B_{max^2}/2\mu_0. \tag{11.9}$$

For example, a laser pointer generating a 3mW beam with a radius of 1 mm has an average intensity of 955 W m^{-2}. Compare this to sunlight, with an average intensity of a little over 1000 W m^{-2}. The peak electric field in the laser beam is $E_{max} = 0.05$ V m^{-1}, and the peak magnetic field is $B_{max} = 1.6 \times 10^{-10}$ T. The energy density, in units of J m^{-3}, is: $u = u_E + u_B = ½\, \varepsilon_0 E^2 + ½\, B^2/\mu_0 = \varepsilon_0 E^2 = B^2/\mu_0$. Exactly half the energy in an electromagnetic wave is in the electric field, and half in the magnetic field. To get from energy density (J m^{-3}) to intensity (J s.m^{-2}), multiply by c (m s^{-1})—see figure 11.1.

11.3 Dimensional homogeneity in electrostatics

Earlier, we illustrated the concept of dimensional homogeneity via mechanics. However, we could also consider a molecule in a strong electric field. This molecule has a permanent dipole moment; that is, the centre of distribution of positive-charge and the centre of the distribution of negative-charge do not coincide. Thus, a molecule such as H–F, has a slightly negatively-charged end (the fluorine atom) and a slightly positively-charged end (the hydrogen atom), and the molecule may be thought of as $^{(\delta-)}$F–H$^{(\delta+)}$. In a molecule such as H_2 or F_2, the centres of the two distributions of charge coincide, and there is no permanent dipole moment.

Let us consider such a molecule with a permanent electric dipole moment, $\mu^{(0)}$, in an intense, non-uniform electric field. The total dipole moment, $\boldsymbol{\mu}$, of the molecule may then be written as

$$\mu_\alpha = \mu^{(0)\alpha} + \alpha_{\alpha\beta}\, E_\beta + ½\beta_{\alpha\beta\gamma}\, E_\beta\, E_\gamma + ⅓A_{\alpha\beta\gamma}E_{\beta\gamma} + \ldots \qquad (11.10)$$

Here, the second term on the RHS gives the dipole moment induced by the applied field through the polarizability of the molecule, and the third term gives the departure from a linear polarization representing the dipole moment induced by the square of the applied field through the first hyperpolarizability of the molecule. $E_{\beta\gamma}$ is a component of the electric field-gradient at our molecule due to the inhomogeneities of the applied field; it will induce a dipole moment through the dipole-quadrupole polarizability of the molecule.[4] The first term on the RHS is the permanent dipole moment; this may be modified by the applied field, because the field will induce dipole moments in the molecule and these smaller induced dipole moments might (depending upon the symmetry) interfere with the magnitude of $\mu^{(0)}$.

As you see, there are many ways of inducing an electric dipole moment in a dipolar molecule. The dimensions and units (in the SI) of these various polarizabilities and related variables are given in table 11.2.

To demonstrate that the electric dipole is being induced via very different electrical properties of the molecule in the electric field, we will rewrite equation (11.10) in dimensional terms. Thus,

[4] I am employing standard Cartesian tensor notation in which a repeated Greek suffix denotes a sum over all three Cartesian components.

Table 11.2. Dimensions of molecular polarizabilities, and related quantities.

Variable	Dimension	SI unit	Description
μ	LTI	Cm (or sA m)	Electric dipole moment
α	$M^{-1} T^4 I^2$	$C^2 m^2 J^{-1}$	Polarizability
β	$M^{-2}L^{-1}T^7I^3$	$C^3 m^3 J^{-2}$	First hyperpolarizability
E	$MLT^{-3}I^{-1}$	$V m^{-1}$	Electric field
E'	$MT^{-3}I^{-1}$	$V m^{-2}$	Electric field-gradient
A	$M^{-1} LT^4I^2$	$C^2 m^3 J^{-1}$	Dipole-quadrupole polarizability

$$[\mu] = LIT = [\alpha][E] = (M^{-1}T^4I^2)(MLT^{-3}I^{-1})$$
$$= [\beta][E]^2 = (M^{-2}L^{-1}T^7I^3)(MLT^{-3}I^{-1})^2 = (M^{-2}L^{-1}T^7I^3)(M^2L^2T^{-6}I^{-2})$$
$$= [A][E'] = (M^{-1}LT^4I^2)(MT^{-3}I^{-1}). \quad (11.11)$$

So, the induced dipole moments are all the same, and they can be added, but the sum of the induced dipoles may well interfere with the magnitude and direction of the permanent dipole moment, μ_0. This is the case, even though the polarizabilities through which the dipole moments are induced are very different in magnitude (α is, of order, 10^{-40} C^2 m^2 J^{-1}, and the dipole-quadrupole polarizability, A is, of order, 10^{-50} C^2 m^3 J^{-1}).

11.4 Molecules and fields

James Clerk Maxwell famously demonstrated that light and electromagnetic phenomena are all parts of a single whole. This is clearly seen from the dimensions of the permittivity of free space and the permeability of free space (see table 11.1). We could use dimensional analysis to identify the product of μ_0 and ε_0. First, assume a relationship between c, μ_0 and ε_0:

$$c = (\varepsilon_0)^a(\mu_0)^b = LT^{-1} = (M^{-1}L^{-3}T^4I^2)^a(MLT^{-2}\,I^{-2})^b$$
$$= M^{-a+b}L^{-3a+b}T^{4a-2b}\,I^{2a-2b}. \quad (11.12)$$

This gives us: $a = b = -½$. So (the constant of nature) $c = 1/\sqrt{\varepsilon_0\mu_0}$, or $c^2 = 1/\varepsilon_0\mu_0$.

We may use the Lorentz force (or electromagnetic force) of a charge (q) moving with velocity, $\boldsymbol{v}$, in an electromagnetic field (that is, a field consisting of both $\boldsymbol{E}$ and $\boldsymbol{B}$) to gain an insight into the dimensions of the phenomena involved in molecular interactions. The expression for the Lorentz force contains two terms:

$$[q(\mathbf{E} + (v \times \mathbf{B}))] = TI.\ MLT^{-3}I^{-1} + TI(LT^{-1}.\ MT^{-2}I^{-1}) = \text{Force} = MLT^{-2}. \quad (11.13)$$

By dimensional homogeneity, the electric field, $\boldsymbol{E}$, must have the same dimensions as $v \times \boldsymbol{B}$. The equation tells us that the electromagnetic force on a charge q is a combination of a force in the direction of the electric field, $\boldsymbol{E}$, proportional to the magnitude of the field, and the magnitude of charge, and a force at right angles to the magnetic field, $\boldsymbol{B}$, and the velocity v of the charge, proportional to the magnitude of the field, the charge and the velocity. Variations on this basic formula describe the magnetic force on a current-carrying wire, the electromotive force in a wire loop moving through a magnetic field (Faraday's law of induction), and the force on a moving charged particle. Figure 11.2 is a representation of the Lorentz force law.

Thus, the Lorentz force law allows us to compute the dimensions of electric field (Vm^{-1} in the SI), velocity of light (ms^{-1} in the SI) and the magnetic flux density (tesla in the SI). Hence,

$$[\boldsymbol{E}] = [v \times \boldsymbol{B}] = \mathsf{MLT^{-3}I^{-1}} = \mathsf{LT^{-1}.\ MT^{-2}I^{-1}}, \tag{11.14}$$

where $\mathsf{MT^{-2}I^{-1}}$ is the dimension of the magnetic flux density (see figure 11.1). The inverse-square law of force between charged particles tells us that if the particle is at rest, $\boldsymbol{B} = 0$ and $\boldsymbol{E} \propto e\,\boldsymbol{r}/r^3$. When the charged particle is in motion with velocity v, $\boldsymbol{B} \propto ev \times \boldsymbol{r}/r^3$. The constants of proportionality needed to give reality to E and B, are generated by a charged particle in the different systems of units. In the rationalised m.k.s., or SI system:

$$\boldsymbol{E} = (e/4\pi\varepsilon_0)(\boldsymbol{r}/r^3), \tag{11.15}$$

and

$$\boldsymbol{B} = (\mu_0 e/4\pi)(v \times \boldsymbol{r})/r^3. \tag{11.16}$$

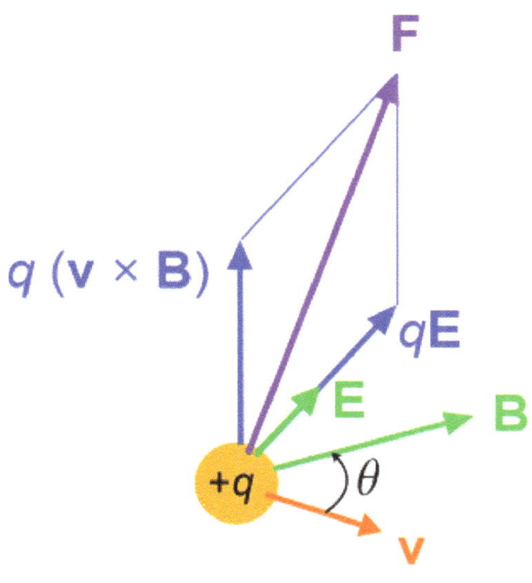

Figure 11.2. The Lorentz force, $\boldsymbol{F}$, on a charged particle (charge $+q$) in motion (instantaneous velocity v). The electric and magnetic fields are time-dependent, and the charge is a scalar. The electric force is $q\boldsymbol{E}$, and the magnetic force is $qvB\sin\theta$. This 'Lorentz force particle' image has been obtained by the author from the Wikimedia website where it was made available under a CC0 1.0 license. It is included within this article on that basis. It is attributed to Maschen.

From equations (11.15) and (11.16) we may derive the dimensions of the E and B fields. Thus,

$$[E] = \mathsf{IT/M^{-1}L^{-3}T^4I^2.\ L^2 = MLT^{-3}I^{-1}}(\mathrm{kg\ ms^{-3}A^{-2}}),\ \text{and}$$
$$[B] = \mathsf{MLT^{-2}I^{-2}.\ IT.\ LT^{-1}/L^2 = MT^{-2}I^{-1}}(\mathrm{kgs^{-2}A^{-1}}).$$

Since a product of $\boldsymbol{B}$ with a velocity is a force per unit charge, the m.k.s. unit of $\boldsymbol{B}$ is that of $\boldsymbol{E}$.m^{-1} sec; that is, Vs m^{-2}. A volt sec is called a weber, and the unit of $\boldsymbol{B}$ is therefore weber m^{-2}, or tesla (see figure 11.1).

11.5 Interacting magnetic dipoles, and the origin of radiation at 21 cm

As an example of the application of electromagnetic interactions, let us look at the origin of the radiation with a wavelength of 21 cm that may be observed in space. We will see that this radiation arises from a magnetic interaction between the positively-charged proton nucleus of the hydrogen atom, with the electron orbiting the nucleus—see figure 11.3 for the details of this interaction. Each component of this interaction (electron and proton) acts like a tiny bar-magnet, whose strength is characterised by their magnetic moments. Whereas an electric dipole is a static manifestation of the separation of charged particles, as in the charge separation in the stable (but reactive) molecule $^{(\delta+)}$H–F$^{(\delta-)}$, a magnetic dipole is different. There is no magnetic monopole, and so a magnetic dipole has dimensions very different from an electric dipole.

An electric dipole, μ, is charge separation by distance; that is, $[\mu] = \mathsf{ITL}$, and has units in the SI of coulomb metre. In a magnetic dipole, the charges are not localised but are in movement, and a magnetic dipole like the magnetic moment of the proton, μ_N, is given by $\mu_N = eh/4\pi m_p$, where e is the protonic charge, h is the Planck

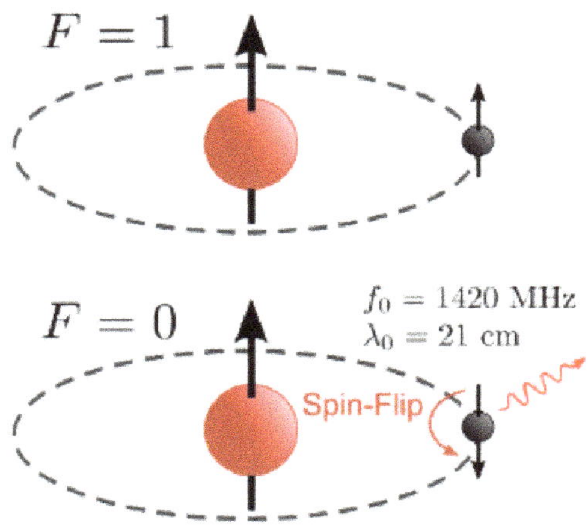

Figure 11.3. Ground state hyperfine levels of the hydrogen atom (upper image: spins parallel and antiparallel in the lower image) with the spin-flip transition, emitting radiation at 1420 MHz (the 21 cm hydrogen line). Reproduced from https://en.wikipedia.org/wiki/Hydrogen_line. Image stated to be in the public domain. It is included within this article on that basis.

constant, and m_p is the mass of the proton. We may find the dimensions of this magnetic moment as $[\mu_N] = [eh/4\pi m_p] = \mathrm{TI.\ ML^2T^{-1}/M = IL^2}$. Thus, a magnetic moment consists of a charge moving continuously over an area—there is no time dependence, because the motion never ceases. If we think of magnetism as arising from circulating charges inside the electron and proton, and simplify the magnetic model to an intrinsic current loop for each particle, then the magnetic moment would be the current going around the loop times its area.

The magnetic moments of the electron and proton have been measured with exceptional precision (electron magnetic moment, $\mu_e = -9.284\,764\,7043(28) \times 10^{-24}$ J T^{-1} and proton magnetic moment, $\mu_p = 1.410\,606\,797\,36(60) \times 10^{-26}$ J T^{-1} [3]), but here we will be content with a rough estimate. The magnitude of the moments is constrained by atomic spectra and the quantum mechanical description of those spectra; that the smallest possible angular momentum associated with the circulating currents is, of order, Planck's constant, h divided by 2π. Imagine, therefore, that the model current loops in the two spins of the hydrogen atom (see figure 11.3) have radii a_e and a_p for the electron and the proton, respectively, that the charges are $-e$ and $+e$ and that the masses of the particle are m_e and m_p. If the entire mass is circulating (as well as the charge) then the magnetic moments, denoted μ_e and μ_p, would be $\mu_e = \pi a_e^2 e/\tau_e c$ and $\mu_p = \pi a_p^2 e/\tau_p c$, and the angular momenta would be $m_e 2\pi a_e^2/\tau_e = h/2\pi$ and $m_p 2\pi a_p^2/\tau_p = h/2\pi$, where $\tau_{e,p}$ are the orbital periods. Here c is the speed of light and is required to obtain the appropriate dimensions for the magnetic moments; for example, $\mu_p = \pi a_p^2 e/\tau_p c$, where $m_p 2\pi a_p^2/\tau_p = h/2\pi$ ($[c\tau_p] = \mathrm{T\ \ LT^{-1} = L}$; $[m_p 2\pi a_p^2/\tau_p] = \mathrm{ML^2T^{-1}} = h$), and so $\mu_p = \tau_p \pi eh/4\pi^2 m_p c\tau_p$. Eliminating $\tau_{e,p}$ from these equations, we have: $\mu_e = eh/4\pi m_e c$ and $\mu_p = eh/4\pi m_p c$.

These relations are correct to an order of magnitude, but because our current loop model oversimplifies the structure of both particles, each must be multiplied by a correction factor, called a g-factor in a fuller treatment. The order of magnitude of the energy of interaction of two magnetic moments may be determined by a simple application of dimensional analysis. We will hypothesise that the energy of interaction (E) of the two spins in the hydrogen atom has the form of a power law,

$$E = (\text{magnetic moment})^a(\text{distance})^b(\text{permeability of the vacuum})^c. \quad (11.17)$$

Dimensionally we write

$$E = \mathrm{ML^2T^{-2}} = (\mathrm{IL^2})^a(\mathrm{L})^b(\mathrm{MLT^{-2}I^{-2}})^c = \mathrm{M}^c\mathrm{L}^{2a+b+c}\mathrm{T}^{-2c}\mathrm{I}^{a-2c}. \quad (11.18)$$

This gives us: $a = 2$, $b = -3$ and $c = 1$. Thus, the simplest interaction of two magnetic dipoles, μ, separated by a distance r is $E(r) = \mu^2.\mu_0/r^3$.

If we consider the two interacting magnetic dipoles to be those of the electron and the proton in a hydrogen atom, $E \sim \mu_e\, \mu_p\, /a_0^3$, where a_0 is the Bohr radius. This quantity may be considered the difference between the energy of a hydrogen atom when the magnetic moments of the proton and the electron point in the same direction and when they point in opposite directions (see figure 11.3). Typically, hydrogen atoms in the interstellar medium will be oscillating between these two states and will be emitting and absorbing electromagnetic radiation with wavelength $\lambda = hc/E$ in the process. Putting in known values of the constants gives radiation with a frequency, of order, 1600 MHz; in reasonable agreement (when we consider the approximations being made) with the more exact result of 21 cm [4].

11.6 Units and the SI

One of the first observations made by students when they begin to study electromagnetism, is the variety of units. These units exist for sound historical reasons [1, 2]. Gaussian units constitute a metric system of physical units, and represent the main alternative to the SI for the description of quantities encountered in electromagnetism. The Gaussian system of units is based on the c.g.s.; that is, centimetre–gram–second base units. The SI system may be called the m.k.s.a. system of units: metre–kilogram–second–ampere (see chapter 1). Today, SI units are predominant in most fields; but they generate huge values for the commonly encountered quantities of electromagnetism, such as electric fields and currents. The SI units of electromagnetism were not designed with molecular physics in mind; they are more appropriate for electrical engineering. Conversions between Gaussian units and SI units are not as simple as with mechanical unit conversions; for example, Maxwell's equations need to be adjusted depending on which system of units one chooses.

With Maxwell's equations, one difference between Gaussian and SI units are the factors of 4π in the various formulae. SI electromagnetic units are termed 'rationalized', because Maxwell's equations have no explicit factors of 4π in the formulae. On the other hand, the inverse-square force laws, such as Coulomb's law and the Biot–Savart law, do have a factor of 4π attached to the distance dependence, r^2. (This quantity of 4π appears because $4\pi r^2$ is the surface area of the sphere of radius r.) In unrationalized Gaussian units (not in Lorentz–Heaviside units) the situation is reversed. Here, two of Maxwell's equations have factors of 4π in the formulae, while both of the inverse-square force laws, Coulomb's law and the Biot–Savart law, have no factor of 4π attached to r^2 in the denominator.

Another major difference between Gaussian and SI units is in the definition of the unit of charge. In the SI, a separate base unit (the ampere) is associated with electromagnetic phenomena; with the consequence that something like electrical charge (1 coulomb = 1 ampere × 1 s) is a unique dimension of a physical quantity, and is not expressed purely in terms of the mechanical units (kilogram, metre, second). On the other hand, in Gaussian units, the unit of electrical charge (the statcoulomb, statC) can be written entirely as a dimensional combination of the mechanical units (gram, centimetre, second); as for example, 1 statC = 1 $\text{g}^{1/2}\,\text{cm}^{3/2}\,\text{s}^{-1}$; Coulomb's law in Gaussian units is, $F = Q_1\, Q_2/r^2$, where F is the force between two electrical charges, Q_1 and Q_2 and r is the distance separating these charges. If Q_1 and Q_2 are expressed in statC and r in cm, then F will be expressed in dyne (a dyne is the unit of force in the c.g.s. system of units; 1 dyn = 1 g cm sec^{-2}; in the SI, force is given in newton with 1 newton (N) = 1 kg m sec^{-2} = 10^5 dyn).

By contrast, Coulomb's law in the SI units is $F = (1/4\pi\varepsilon_0)\, Q_1\, Q_2\, /r^2$; where ε_0 is the vacuum permittivity, a quantity with a dimension (see table 11.2). Without ε_0, the two sides would not have consistent dimensions in the SI; in fact, the quantity ε_0 does not exist in Gaussian units. This is an example of how some dimensional physical constants can be eliminated from the expressions of physical laws simply by the judicious choice of units. In the SI, $1/\varepsilon_0$, converts or scales flux density, **D**, to electric field, **E** (the latter has dimension of force per charge), while in rationalized Gaussian units, flux density is the same as electric field in free space (see figure 11.2).

Since the unit of charge is built out of mechanical units (mass, length, time), the relation between mechanical units and electromagnetic phenomena is clearer in Gaussian units than in the SI. In particular, in Gaussian units, the speed of light c shows up directly in electromagnetic formulae such as Maxwell's equations, whereas in the SI it only shows up implicitly via the relation, $\mu_0\varepsilon_0 = 1/c^2$. In Gaussian units, unlike SI units, the electric field $\boldsymbol{E}$ and the magnetic field $\boldsymbol{B}$ have the same dimensions. This amounts to a factor of c difference between how $\boldsymbol{B}$ is defined in the two systems of units. The same factor applies to other magnetic quantities such as $\boldsymbol{H}$ and $\boldsymbol{M}$; for example, in a planar light wave in vacuum, $|\mathbf{E}(\boldsymbol{r}, t)| = |\mathbf{B}(\boldsymbol{r}, t)|$ in Gaussian units, while $|\mathbf{E}(\boldsymbol{r}, t)| = c|\mathbf{B}(\boldsymbol{r}, t)|$ in SI units.

Looking at interacting charges *in vacuo* is all very well for a discussion of forces, but electromagnetic phenomena are encountered in the laboratory in electrical circuits. And the study of the magnetic fields of currents in wires and conductors has played, and still does play, a central role in electromagnetism. In particular, it happens that the practical electrical units that we use every day stem from a definition of the unit of current (see table 1.3), which prior to May 2019 was based on the force between two line-currents; that is, two infinite, parallel wires carrying electrical currents. This definition is impossible to realise in the laboratory, and served to fix the value of the permeability of free space, μ_0. Since May 2019, the definition of the ampere is *The elementary charge e is 1.602 176 634 × 10^{-19} C*. Thus, an amount of current is now defined by counting electrons passing along a wire. And the permeability of free space is no longer a constant of Nature, but is now an experimental observable with an associated uncertainty (see table 1.3) [2].

In the old definition of the ampere (table 1.3); two rectilinear current lines, I and I ′, are parallel at a separation r. The magnetic field of one current acts on the moving charged particles of the other (and vice versa). The force per unit length, F/l, is given by the current law of Ampère:

$$\frac{F}{l} = k_{\mathrm{A}}\left\{\frac{I.\,I'}{2\pi r}\right\}, \tag{11.19}$$

where k_{A} is the magnetic force constant from the Biot–Savart law, F/l is the total force on either wire per unit length of the shorter (the longer is approximated as infinitely long relative to the shorter), r is the distance between the two wires, and I and I' are the direct currents carried by the wires. This is a good approximation if one wire is sufficiently longer than the other; it can then be approximated as being infinitely long, and if the distance between the wires is small compared to their lengths (so the one infinite-wire approximation holds), but large compared to their diameters (so that they may also be approximated as infinitely thin lines). The value of k_{A} depends upon the system of units chosen, and the value of k_{A} decides how large the unit of current will be in that system of units. The force could be attractive or repulsive, depending upon the relative phases of the two line-currents. Before May 2019, k_A was equal to $\mu_0/4\pi$, by definition. Thus, a current of 1 amp (0.1 abamp), passing in opposite directions through the wires with $r = 1$ m, generated a force between the two line-currents of 2×10^{-7} newton per metre of length.

The SI unit of charge is the coulomb, that is, 1 A s (amp second), which is the basis of the present definition of the ampere in the new Quantum-SI (see table 1.3). A current of 1 ampere in a wire carries charge along that wire at a rate of 1 C per second. Before the redefinition of the SI in 2019, the value of the electron charge was about 1.602×10^{-19} C; it was a measurable quantity with an associated uncertainty that depended upon the experimental procedure [2]. Since May 2019, the value of the electron charge has been fixed at a value of $1.602\ 176\ 634 \times 10^{-19}$ C. A coulomb represents the passage of about 6.3×10^{18} electrons, and a current of 1 amp in a wire represents the passage of this number of electrons per second. If N is the number of particles per unit length of line, v their velocity and e the charge on each particle, the current I is given by $I = Nev$.

11.7 Final point: electro- and magneto-optics

To identify which combination of the electromagnetic variables will appear in commonly encountered phenomena and experiments, we need to identify the combination of variables that has dimensions of energy ($\mathsf{ML^2T^{-2}}$). This is an ideal use of dimensional analysis. We begin by saying that in electro-optic phenomena, the energy of the interaction between the applied field and the molecule is: Energy $\propto$ (polarizability)^a (electric field)^b $(\varepsilon_0)^c$, and in magneto-optic phenomena it is: Energy $\propto$ (magnetic susceptibility)^d (magnetic field)^e $(\mu_0)^f$; that is, we are seeking a combination of the variables with the dimensions of energy. The dimensional equations are

$$\mathsf{ML^2T^{-2}} = (\mathsf{M^{-1}T^4I^2})^a(\mathsf{MLT^{-3}I^{-1}})^b(\mathsf{M^{-1}L^{-3}T^4I^2})^c, \text{ and}$$

$$\mathsf{ML^2T^{-2}} = (\mathsf{M^{-1}L^2T^2I^2})^d(\mathsf{L^{-1}I})^e(\mathsf{MLT^{-2}I^{-2}})^f, \text{ respectively.} \qquad (11.20)$$

This generates

$$\mathsf{ML^2T^{-2}} = \mathsf{M}^{-a+b-c}\mathsf{L}^{b-3c}\mathsf{T}^{4a-3b+4c}\mathsf{I}^{2a-b+2c} = \mathsf{M}^{-d+f}\mathsf{L}^{2d-e+f}\mathsf{T}^{2d-2f}\mathsf{I}^{2d+e-2f}. \qquad (11.21)$$

The solutions are: $a = 1$, $b = 2$, $c = 0$, $d = 1$, $e = 2$ and $f = 2$.

Thus, the combination of these variables having the dimension of energy is αE^2 and $\chi H^2\mu_0^2 = \chi B^2$. To form dimensionless combinations of these electromagnetic variables (in the SI) to investigate macroscopic phenomena described by, for example, the Debye equation of polarization, the Kerr effect and the Cotton–Mouton effect, we would have to find the nondimensional quantity:

$$\frac{[\text{Energy of the electro magnetic interaction}]}{[\text{Available thermal energy}]} = \frac{\alpha E^2}{k_B T} = \frac{\chi B^2}{k_B T} \qquad (11.22)$$

In all such electro-optic experiments, one is seeking to impose an order on the fluid sample by application of: a static electric field (Kerr effect), an optical electric field (optical Kerr effect), a perpendicular magnetic field (the Cotton–Mouton effect), or an applied electric field-gradient (the Buckingham effect to measure molecular quadrupole moments). This imposed order is opposed by the random, thermal motion of the molecules in the sample. The applied electromagnetic field turns the fluid into a uniaxial crystal, and induces a small birefringence. If the applied field is modulated at some frequency, and the signal at that modulated frequency is identified, then it is possible to distinguish between the small induced effect and the large noisy, thermal background. But the observable in such electro-optic and magneto-optic experiments is the dimensionless ratio of some interaction energy of the molecule with an applied field, with the thermal energy of the sample.

One last point on induced electric and magnetic dipoles concerns the optical activity tensor, $\boldsymbol{G}'$, which is only non-zero in molecules of very low symmetry (those that display natural optical activity). This molecular property defines the magnetic dipole induced in a molecule by an applied electric field, and the electric dipole induced in the same molecule by an applied magnetic field; that is, $m = G'E$ and $\mu = G'B = G'\mu_0 H$. The dimension of this curious property is:

$$[G'] = \frac{[m]}{[E]} = \mathsf{L^2I/MLT^{-3}I^{-1}} = \frac{[\mu]}{[B]} = \frac{[\mu]}{[\mu 0H]} = \mathsf{LTI/MT^{-2}I^{-1}} = \mathsf{M^{-1}LT^3I^2}. \tag{11.23}$$

A perfect illustration of how to find the dimensions of a little-known, exotic phenomenon.

References

[1] *An Introduction to Electromagnetic Theory* ed Clemmow P C 1973 (Cambridge: Cambridge University Press)

[2] Williams J H 2020 *Defining and Measuring Nature: The Make of All Things* 2nd edn (Bristol: IOP Publishing)

[3] Fundamental Physical Constants | NIST https://www.nist.gov/pml/fundamental-physical-constants

[4] Ewen H I and Purcell E M 1951 Observation of a line in the Galactic radio spectrum: radiation from galactic hydrogen at 1420 Mc/sec *Nature* **168** 356

IOP Publishing

Dimensional Analysis
The great principle of similitude
Jeffrey H Williams

Chapter 12

The dynamics of atoms and molecules

'Truth is found neither in the thesis nor the antithesis, but in an emergent synthesis which reconciles the two'.

Georg Wilhelm Friedrich Hegel (1770–1831)

By the early-1920s, the molecule meme, that is, the acceptance of the view that atoms, and molecules are the basis of solid matter had become widely accepted by the physics community. This conversion of the physics community to what chemists already knew led to the creation of a new area of chemistry—physical chemistry. And it was via physical chemistry, and its transformation into chemical physics that chemistry and physics have grown closer together. This merging was brought about by the development of spectroscopy and the use of quantum mechanics to analyse the dynamics of atoms and molecules.

12.1 Rutherford's model of the hydrogen atom

A table, an iceberg, or our bodies all contain a great deal of empty space between the constituent atoms. Yet these materials are all deemed to be solid and not easily compressible. This paradox in the nature of matter has long been of interest. Indeed, it was an experiment performed in the laboratory of Ernest Rutherford (1871–1937) which fundamentally changed our appreciation of what we term 'solid matter'. Johannes Geiger (1882–1945) worked with Ernest Rutherford at the University of Manchester, and in 1909, along with a young New Zealand student Ernest Marsden (1889–1970), undertook one of the most celebrated of experiments in atomic science: the Geiger–Marsden, or gold-foil experiment. The two young researchers had been told by Rutherford to investigate the transmission of a thin foil of gold by a form of radiation, which today we term alpha-particles. At that time, however, this radiation was known only to consist of a stream of heavy particles that carried a positive

doi:10.1088/978-0-7503-3655-0ch012

electrical charge. Alpha-particles are, in fact, doubly-ionized helium atoms, He^{2+}; a bare helium nucleus consisting of two protons (carrying the positive charge) and two (electrically neutral) neutrons, but these details were not known at that time.

Rutherford wished to probe the nature of the atoms comprising the gold foil. At that time, the model of the atom held that the nucleus contained protons, with electrons to balance the electrical charges, and that there were other electrons moving around this composite nucleus.

This was the 'plum-pudding' model of the atom, developed in 1904 by Joseph John 'J J' Thomson. The Geiger–Marsden experiment consisted of a series of measurements to determine the facility of transmission of the alpha-particles by the foil. If Thomson's plum-pudding model was correct, the heavy alpha-particles should pass through the gold foil with only a few deflections. Although most of the alpha-particles did indeed pass through the gold foil, others were deflected at small angles, and some (about one in every eight thousand) were reflected back to the source of the alpha-particles. This result demonstrated that the plum-pudding model of the atom was incorrect. The fact that some of the positively-charged alpha-particles were reflected meant that the scattering atoms had a concentrated centre of positive electrical charge, and were of high mass. The reflected alpha-particles had either hit the positively-charged centre directly, or passed by it close enough to be affected by its positive charge (by electrostatic repulsion). Since many other alpha-particles passed through the gold foil, the positive centre would have to be a small size compared to the rest of the atom. Because the majority of the alpha-particles continued on their original path through the gold foil, Rutherford concluded that most of the remainder of the atom was a region of very low density. This model of the atom remains the current view of the atom's structure: a positively-charged centre, or nucleus of high density and negatively-charged electrons moving around the nucleus at relatively large distances compared to the nuclear radius (like a tiny solar system). The atom was seen to consist of a small number of interacting electrical charges in a large volume of vacuum.

In this scattering experiment, Rutherford had to consider how close the beam of alpha-particles approached the nuclei of the gold atoms. The essential quantity needed to examine the scattering event is the differential cross-section, σ (of dimensions L^2), defined as the number of scattered particles per unit time in a solid angle, and divided by the incident intensity of the alpha-particles' beam [1]. In order to be able to compute σ, Rutherford needed to find a relation between the impact parameter b (of dimension L) of a single alpha-particle and the scattering angle θ (See figure 12.1 for details); σ depends on the impact parameter and the scattering angle as follows:

$$\sigma = \frac{b}{\sin\,\theta}\left|\frac{db}{d\theta}\right|. \tag{12.1}$$

Using the orbit equation of a particle under the influence of a central force, such as the electrostatic Coulomb interaction, it is possible to obtain $b = b(\theta)$. Following this procedure, the desired relation is:

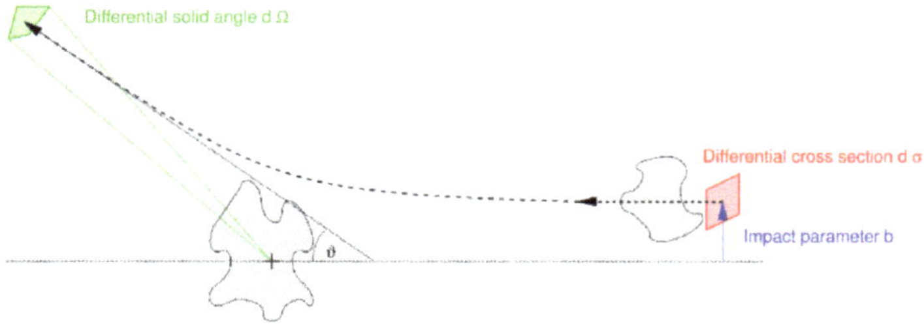

Figure 12.1. Scheme of the Geiger–Marsden experiment. An alpha-particle (the scattering particle) passes close to the nucleus of the gold atom (scattering centre); due to the strong electrostatic interaction it is scattered, and changes its trajectory by an angle θ. This 'differential cross section' image has been obtained by the author from the Wikimedia website where it was made available under a CC BY-SA 3.0 license. It is included within this article on that basis. It is attributed to Vswitchs.

Table 12.1. Variables for Rutherford scattering.

Variable	Dimension	Description
v	LT^{-1}	Velocity of alpha-particle
b	L	Scattering length, or impact parameter
θ		Scattering angle
m_a	M	Mass of scattering particle
F_e	MLT^{-2}	Interaction force between the heavy nucleus and the incident alpha-particle

$$b = \{kZZ'e^2/2E\}\cot\frac{\theta}{2}. \tag{12.2}$$

Here Z and Z' are the atomic numbers of the heavy nucleus and the incident alpha-particle, respectively, k is the force constant of the electrostatic Coulomb interaction ($k = 1/4\pi\varepsilon_0$), $E = ½m_\alpha v^2$ is the kinetic energy of the alpha-particle and e is the electron charge.

Let us now consider how well dimensional analysis can explain the scattering, which so surprised Rutherford. We first choose the set of fundamental variables of the problem of the scattering of alpha-particles off heavy nuclei, as v, b, θ, m_α and F_e; these variables are defined in table 12.1.

In this problem there are five variables containing the three mechanical dimensions, and for the Π-theorem we choose v, b and F_e to be the repeating variables. The Π-theorem tells us that there are only two dimensionless parameters to be identified; the first is the scattering angle itself since this is already a nondimensional quantity; that is, $\Pi_1 = \theta$, and the second Π-group can be found using the three repeating variables and the mass of the alpha-particle, m_α:

$$\Pi^2 = m_a(v)^a(b)^b(F_e)^c = \mathsf{M}^{1+c}\mathsf{L}^{a+b+c}\mathsf{T}^{-a-2c}. \tag{12.3}$$

The solutions are: $a = 2$, $b = -1$ and $c = -1$, and we may write: $\Pi^2 = m_\alpha v^2/bF_e$.

The electric force may be written as: $F_e = kZZ'e^2/b^2$, and inserting this into the expression for Π_2 gives: $\Pi_2 = m_\alpha v^2 b/kZZ'e^2$. The Π-theorem allows us to write $\Pi_2 = f(\Pi_1)$, and we obtain an expression for b:

$$b = \{kZZ'e^2/m_\alpha v^2\}f(\theta), \tag{12.4}$$

which is in agreement with equation (12.2) derived from a full analysis [1]—except for the numerical factors.

12.2 The earliest quantum view of the atom

Ernest Rutherford's model of the hydrogen atom caused a considerable stir in physics. According to classical electrodynamics, the electron orbiting the nucleus in Rutherford's model in being accelerated towards the nucleus of the atom by the centripetal force, should:

- emit radiation over a continuous range of frequencies with power proportional to the square of its acceleration, and
- lose potential energy and gain kinetic energy as it spirals into the nucleus, leading to the collapse of the atom.

This does not happen, and the first prediction is contradicted by experiment, as detailed studies of the absorption and emission spectra of the hydrogen atoms became available. Johann Balmer showed in 1885 that the radiation emitted by the excited hydrogen atom consists of discrete frequencies. The spectral lines of radiation from the hydrogen atom were found to fit an empirical (without theoretical foundation) rule—the Balmer–Rydberg formula,

$$\nu_{nq} = 1/\lambda_{nq} = R(1/n^2 - 1/q^2), \tag{12.5}$$

where λ_{nq} is the wavelength, ν_{nq} the wavenumber, n and q are integers with $q > n$ and R is the Rydberg constant. The constant, R, was found by measurements to be 1.097×10^7 per metre. Equation (12.5) gives the spectral series limit ($n \to \infty$) as $n_n = R/n^2$. These observations baffled physicists at the end of the 19th Century. Then, the Danish physicist, Niels Bohr (1885–1962), in a splendid display of original thought, rescued the hydrogen atom from radiating and collapsing. He derived the Balmer–Rydberg formula for the spectral lines of radiation from the hydrogen atom by invoking the new quantum theory and proposing two postulates [2]:

- The electron, in Rutherford's model of the hydrogen atom, can revolve, without radiating, around the nucleus in allowed quantum, or stable orbits for which the angular momentum L_n is quantized; such that:

  $$L_n = nh/2\pi, \tag{12.6}$$

  where n, the quantum number, is an integer greater than zero, and h is the Plank constant.

- An excited electron 'moves' from a stable orbit of radius r_q, corresponding to quantum number q and total energy (kinetic and potential) E_q, to an inner orbit of radius r_n, corresponding to quantum number n and total energy E_n. The electron loses potential energy and gains kinetic energy and, in the process, it emits radiation of frequency, ν_{nq}, in accordance with de Broglie's hypothesis [3], such that:

$$E_q - E_n = h\nu_{nq}, \tag{12.7}$$

where n is a number greater than zero but less than q.

If we apply the two postulates of Bohr to the hydrogen atom, whose electron of mass, m_e and charge, $-e$ revolves with velocity v_n about a stationary nucleus of mass, M and charge, $+e$ in a circular orbit of radius r_n, the angular momentum of the electron, in this nth orbit, is:

$$L_n = \frac{nh}{2\pi} = m_e v_n r_n. \tag{12.8}$$

Niels Bohr proceeded by applying classical mechanics to calculate the orbits of the electron around the nucleus. Bohr stated that the electron is held in a circular orbit by electrostatic attraction, and that by dimensional homogeneity one is able to equate two forces: the electrostatic Coulomb force and the centripetal force:

$$m_e v_n^2/r_n = e^2/4\pi\varepsilon_0 r_n^2. \tag{12.9}$$

Equations (12.8) and (12.9) give the speed and radius of revolution as:

$$v_n^2 = e^2/2\varepsilon_0 nh, \text{ and } r_n = \varepsilon_0 n^2h^2/\pi m_e e^2. \tag{12.10}$$

Thus, we have the velocity of the electron, at any radius (in any orbit). It also determines the total energy of the electron at any radius, which is negative and inversely proportional to r. Thus, it requires energy to pull the orbiting electron away from the nucleus (to ionize the atom). With $n = 1$, we have the Bohr radius (a_0), which is today known rather precisely: $r_1 = a_0 = 5.291\ 772\ 109\ 03(80) \times 10^{-11}$ m. The total energy (kinetic and potential) in the nth quantum state is:

$$E_n = ½m_e v_n^2 - e^2/4\pi\varepsilon_0 r_n. \tag{12.11}$$

Substituting for v_n and r_n from equation (12.10) gives:

$$E_n = m_e e^4/8\varepsilon_0^2 n^2 h^2 \approx -13.6/n^2\text{eV}. \tag{12.12}$$

An electron in the lowest energy level of hydrogen ($n = 1$) therefore has about 13.6 eV less energy than a motionless electron infinitely far from the nucleus. The next energy level ($n = 2$) is -3.4 eV. The third ($n = 3$) is -1.51 eV, and so on. The total energy in the qth quantum state is:

$$E_q = m_e e^4/8\varepsilon_0^2 q^2 h^2. \tag{12.13}$$

Equation (12.5) then becomes:

$$E_q - E_n = m_e e^4/8\varepsilon_0{}^2h^2\{1/n^2 - 1/q^2\} = h\nu_{nq}, \tag{12.14}$$

$$\text{where } \nu_{nq}/c = 1/\lambda_{nq} = m_e e^4/8c\varepsilon_0{}^2h^3\{1/n^2 - 1/q^2\}. \tag{12.15}$$

The Rydberg constant, R, in equation (12.5), is given by:

$$R = m_e e^4/8c\varepsilon_0{}^2h^3. \tag{12.16}$$

Substituting values of the physical constants into equation (12.16), R is found as 10 973 731.568 160(21) m^{-1}. This expression contains a number of terms: $m_e c^2$ is the rest mass energy of the electron (~511 keV); and $e^2/(4\pi\varepsilon_0)c\hbar = \mu_0 ce^2/2h = \alpha \approx 1/137$ is the fine-structure constant.

The stabilization of Rutherford's nuclear model of the hydrogen atom by Niels Bohr was recognised as a major advance. It gave an additional impetus to the development of quantum mechanics. However, the limitations of this model quickly became apparent by subsequent experiment. Let us now look at the mechanics of the simplest atom via dimensional analysis (see table 12.2 for details of variables).

If an electron is moving in an orbit with period T, classically, electromagnetic radiation will be radiated to every orbital period over a range of frequencies. In quantum mechanics, this emission must be in quanta of light, of frequencies consisting of integer multiples of $1/T$. This means that the energy level corresponding to a classical orbit (n) of period $1/T$ must have nearby energy levels that differ in energy by $\hbar/T = \hbar\nu$, and they should be equally spaced near the energy of the level of interest, E_n; the energy spacing will be: $\Delta E = \hbar\nu_{n\to n'}$; where the transition is from state n to state, n', and the transition frequency is $\nu_{n\to n'}$.

Bohr considered that when an electron was moving in an orbit, it could be analysed by classical mechanics. It was only when a transition to another state, and another orbit occurred, that this classical approach had to be modified. The level spacing between classical orbits can be calculated from what became known as the correspondence principle. The hydrogen atom is a Kepler problem; it comprises two charged particles—one of which is much heavier than the other, and can be

Table 12.2. Variables for the hydrogen atom.

Variable	Dimension	Description
m_e	M	Mass of the electron
e	IT	Charge of the electron
M	M	Mass of the nucleus
v_n	LT^{-1}	Velocity of electron in state n
r_n	L	Radius of orbit of state n
ε_0	$M^{-1}L^{-3}T^4I^2$	Permittivity of free space
h	ML^2T^{-1}	Planck constant
c	LT^{-1}	Speed of light

considered immobile—interacting by Coulomb's law of electrostatics. This is another inverse square central force. For a hydrogen atom, the classical orbits have a period T determined by Kepler's third law (see section 6.2) scaling as $r^{3/2}$; so T^2 scales as r^3. The energy scales as $1/r$, so the level spacing becomes: $\Delta E \propto 1/r^{3/2} \propto E^{3/2.}$ The angular momentum, L, of a circular orbit scales as $\sqrt{r}$; the energy, in terms of angular momentum is: $E \propto 1/r \propto 1/L^2$. Bohr assumed that quantized values of L are equally spaced, the spacing between neighbouring levels is then: $\Delta E \propto 1/(L + \hbar)^2 - 1/L^2 \approx -2\hbar/L^3 \propto -E^{3/2}$. The spacing will be $\hbar$; so, the angular momentum in the hydrogen atom should be an integer multiple of $\hbar$: $L = n\hbar$.

Applying dimensional analysis (the Rayleigh algorithm) to the hydrogen atom, that is, an electron orbiting a proton, we may hypothesize that the possible variables that determine the size (radius of an orbit, r) of this atomic system are its properties such as: the charge, e; a measure of the strength of the electrical force between the proton and electron, ε_0; the mass of the electron, m_e. The much heavier nucleus is assumed to be stationary. If we include only these three parameters; that is, assume: $r \propto (m_e)^x\ (e)^y\ (\varepsilon_0)^z$ we will see that the resulting system of equations for the values of the exponents is overdetermined; that is, there are more equations than unknowns. An overdetermined system is almost always inconsistent—it has no unique solution. However, if you include a fourth quantity in the analysis, h, the Planck constant, then we will find that the problem is now solvable. That a solution to this problem requires the inclusion of h, means that the size of the atom is quantum mechanical in nature, and that without quantum mechanics, atoms would not have a well-defined size. In other words, it implies that quantum physics must play an important role in atomic structure. Consequently, we now assume that the atomic radius r can be expressed as a product (see table 12.2 for definitions),

$$r = (e)^a(m_e)^b(\varepsilon_0)^c(h)^d. \tag{12.17}$$

The dimensional equation is

$$\mathsf{L} = (\mathsf{IT})^a(\mathsf{M})^b(\mathsf{M}^{-1}\mathsf{L}^{-3}\mathsf{T}^4\mathsf{I}^2)^c(\mathsf{ML}^2\mathsf{T}^{-1})^d = \mathsf{M}^{b-c+d}\mathsf{L}^{-3c+2d}\mathsf{T}^{a+4c-d}\mathsf{I}^{a+2c}. \tag{12.18}$$

The solutions are: $a = -2$, $b = -1$, $c = 1$ and $d = 2$. Therefore, the formula for the radius of the hydrogen atom is: $r = C\,\varepsilon_0 h^2/m_e e^2$. As is the case with all dimensional analysis, there is no way to determine the dimensionless numerical factor, C, but $C = 1/\pi$, when r equals the Bohr radius (see equation (12.10)). The expression for the dimension in equation (12.18) requires $a = -2c$, which means that e and ε_0 always occur as e^2/ε_0. This makes sense, because they both determine the strength of the electric force between two particles of charge e.

What about relativistic effects? Should we have included the speed of light, c, in our analysis? Is the electron speeding around its orbit at relativistic velocities? To investigate this idea, we have to add a new term, the speed of light, c, to our initial hypothesis for r,

$$r = (e)^a(m_e)^b(\varepsilon_0)^c(h)^d(c)^e. \tag{12.19}$$

Equating dimensions now gives

$$\begin{aligned} \mathsf{L} &= (\mathsf{IT})^a(\mathsf{M})^b(\mathsf{M}^{-1}\mathsf{L}^{-3}\mathsf{T}^4\mathsf{I}^2)^c(\mathsf{ML}^2\mathsf{T}^{-1})^d(\mathsf{LT}^{-1})^e \\ &= \mathsf{M}^{b-c+d}\mathsf{L}^{-3c+2d+e}\mathsf{T}^{a+4c-d-e}\mathsf{I}^{a+2c}. \end{aligned} \tag{12.20}$$

Now, however, we have an underdetermined system; there is insufficient information to solve for all five exponents. A more likely solution to the problem of introducing relativistic effects is to assume that the radius is

$$r = (e)^a(m_e)^b(\varepsilon_0)^c(h)^d f\left(\frac{v}{c}\right), \tag{12.21}$$

where f is a dimensionless function of velocity of electron/c. This will result in an expression for r multiplied by function f; and other considerations (physical insight and/or experiments) will be needed to determine the functional form of f.

In the early days of quantum mechanics—before the Dirac-Heisenberg synthesis of quantum mechanics—a quantity called the 'classical electron radius' caused much concern. After all, if you could determine the radius of a quantum particles, how did this fit in with the uncertainty principle. At this time, the classical electron radius, r_e, was defined to be $r_e = e^2/4\pi\varepsilon_0\, m_e c^2 \approx 2.8 \times 10^{-15}$ m, where c is the speed of light. This expression was obtained by setting the energy necessary to assemble the electron's charge e equal to $m_e c^2$; the idea being that the energy of assembly went toward the rest-energy of the electron. Experimentally, however, the 'radius' of the electron was found to be less than 10^{-22} m (by scattering experiments). However, this classical radius (classical here meaning not quantum) is a useful quantity that describes some of the short-range interactions of the electron. Dimensional analysis can be used to determine this radius. Consider a sphere of electric charge e, and radius r. What quantities might its mass depend upon? Since the work necessary to assemble the charge depends on the electric force, the mass should depend on e, r and also ε_0, which determines the strength of the electric force—how it propagates through space. If we set the electron mass equal to the product

$$m_e = (e)^a(r)^b(\varepsilon_0)^c \rightarrow \mathsf{M} = \mathsf{M}^{-c}\mathsf{L}^{b-3c}\mathsf{T}^{a+4c}\mathsf{I}^{a+2c}. \tag{12.22}$$

The system is again clearly overdetermined—the expressions for T and I are inconsistent. We originally hypothesised that the sphere of charge was empty space, and so the only variables of interest were those that dealt with the electrical force. But empty space does have properties, in the sense that the electric force is carried by photons, which travel at the speed of light. Therefore, we should include c in our list of possible dependencies (the photons have to be exchanged across the sphere—as in retardation effects in dispersion forces), and we must revise our initial hypothesis (equation (12.22)) to be

$$m_e = (e)^a(r)^b(\varepsilon_0)^c(c)^d. \tag{12.23}$$

This gives us the dimensional equation

$$\mathsf{M} = (\mathsf{IT})^a(\mathsf{L})^b(\mathsf{M}^{-1}\mathsf{L}^{-3}\mathsf{T}^4\mathsf{I}^2)^c(\mathsf{LT}^{-1})^d = \mathsf{M}^{-c}\mathsf{L}^{b-3c+d}\mathsf{T}^{a+4c-d}\mathsf{I}^{a+2c}. \quad (12.24)$$

This time, the unique solutions are: $a = 2$, $b = -1$, $c = -1$ and $d = -2$, which gives $m_e = D\ (e^2/\varepsilon_0 rc^2)$, where D is an unknown dimensionless constant. This result is identical to the result given above for the classical electron radius—except for the numerical factors.

12.3 Electric dipole transitions

The basis of all spectroscopy is the coupling of the oscillating electric field, $\mathcal{E}(t)$, of the incident radiation to a dipole moment within the molecule in the radiation field. Spectroscopists speak of electric dipole allowed transitions, and electric dipole forbidden transitions. The difference between these two types of transitions is magnitude, or intensity. An electric dipole forbidden transition may occur due to the presence of a perturbation, but it will always be much weaker in intensity than an electric dipole allowed transition.

Consider a molecule exposed to radiation with its electric vector lying in the z-direction, and oscillating at a frequency, $\omega = 2\pi\nu$. This field will perturb the charge distribution of the molecule; this perturbation is $H = -\mu_z\,\mathcal{E}(t)$, where $\mathcal{E}(t) = 2\mathcal{E}_0\cos\omega t$. The transition rate ($W_{i\to f}$ with dimension, T^{-1}) from an initial state, $|i\rangle$ to a continuum of final states $|f\rangle$, due to a perturbation of this form is given by the Fermi Golden Rule [4]: $W_{i\to f} = 2\pi\hbar|H|^2\rho(E_{fi})$, and in this instance is $W_{i\to f} = \frac{2\pi}{\hbar}|\mu_{z,fi}|^2\,\mathcal{E}^2\rho(E_{fi})$, where $\rho(E_{fi})$ is the density of the continuum states at an energy $E_{fi} = \hbar\omega_{fi}$, with ω_{fi} being the transition frequency. In a fluid sample, the z-direction corresponds to all possible directions in the molecules, so we replace $|\mu_{z,fi}|^2$ by its mean value $⅓|\mu_{fi}|^2$; that is, the vector becomes a scalar, but the dimensions are unchanged.

The energy of a classical electromagnetic field, which is driving an absorption, or stimulating an emission is $E = ½\int\{\varepsilon_0 \langle\mathcal{E}^2\rangle + \mu_0 \langle H^2\rangle\}\mathrm{d}\tau$, where $\langle\mathcal{E}^2\rangle$ and $\langle H^2\rangle$ are the time-averages of the squared field strengths (electric and magnetic, respectively) and $\mathrm{d}\tau$ is the volume element. In the present case, because the period is $2\pi/\omega$, $\langle\mathcal{E}^2\rangle = 4\mathcal{E}^2/2\pi/\omega \int_0^{2\pi/\omega}\cos^2\omega t\ \mathrm{d}\tau = 2\mathcal{E}^2$. From electromagnetism, $\varepsilon_0 \langle\mathcal{E}^2\rangle = \mu_0 \langle H^2\rangle$. So, for a field in a volume, v, the energy content, $E = 2\varepsilon_0\,\mathcal{E}^2 V$, or $\mathcal{E}^2 = E/2\varepsilon_0 V$; that is, dimensionally, $\mathsf{ML}^2\mathsf{T}^{-2} = \mathsf{M}^{-1}\mathsf{L}^{-3}\mathsf{T}^4\mathsf{I}^2.\ (\mathsf{MLT}^{-3}\mathsf{I}^{-1})^2\,\mathsf{L}^3$, and for the magnetic field, $\mathsf{ML}^2\mathsf{T}^{-2} = \mathsf{M\ \ LT}^{-2}\mathsf{I}^{-2}.\ (\mathsf{L}^{-1}\mathsf{I})^2\,\mathsf{L}^3$. It follows that $W_{i\to f} = (\pi/3\hbar\varepsilon_0)\ |\mu_{fi}|^2\ \{E\,\rho(E_{fi})/V\}$. This expression is for the transition rate from an initial state $|i\rangle$ to a continuum of states $|f\rangle$. In spectroscopy we are interested in the transition rate from a discrete initial state $|i\rangle$ to a discrete final state $|f\rangle$. We now need to introduce the density of radiation states, $\rho'_{\mathrm{rad}}(E)$, where $\rho'_{\mathrm{rad}}(E)\mathrm{d}E$ is the number of waves with photon energies in the range E to $E + \mathrm{d}E$. This results in $W_{i\to f} = (6\varepsilon_0\hbar^2)^{-1}|\mu_{fi}|^2\ \{E_{fi}\,\rho'_{\mathrm{rad}}(E_{fi})/V\}$.

We note that the term $E_{fi}\,\rho'_{\mathrm{rad}}(E_{fi})/V$ is the product of the energy of a monochromatic wave and the density of radiation states divided by the volume (dimensionally, we would write: $\mathsf{ML}^2\mathsf{T}^{-2}\,\mathsf{L}^{-3}\,\rho'_{\mathrm{rad}}(E_{fi}) = \mathsf{ML}^{-1}\mathsf{T}^{-2}\,\rho'_{\mathrm{rad}}(E_{fi})$); hence it is the energy density of radiation states, and we write it as $\rho_{\text{rad}}(E)$. Therefore, $W_{i\to f} = (6\varepsilon_0\hbar^2)^{-1}|\mu_{fi}|^2\ \rho_{\text{rad}}(E_{fi})$. and we can identify the coefficient of stimulated absorption as

$$B = |\mu_{fi}|^2/6\varepsilon_0\hbar^2. \tag{12.25}$$

Thus, $W_{i\to f} = B\ \rho_{\text{rad}}(E_{fi})$. The energy density per unit frequency interval has dimensions: $\mathsf{ML^2T^{-2}/L^3\ T^{-1} = ML^{-1}T^{-1}}$, and B will have dimensions of (volume × frequency)/(energy × time) = $\mathsf{L^3.T^{-1}/ML^2T^{-2}\ T}$. So, $[W_{i\to f}] = \mathsf{T^{-1}} = [B]$ $[\rho_{\text{rad}}(E_{fi})] = \mathsf{M^{-1}L\ \ ML^{-1}T^{-1}}$.

It then follows (see Atkins for full details [4]) that the coefficient of spontaneous emission is:

$$A = 8\pi h(v_{fi}/c)^3(6\varepsilon_0\hbar^2)^{-1}\left|\mu_{fi}\right|^2 = 8\pi h(v_{fi}/c)^3 B = \{8\pi^2 v_{fi}{}^3/3\varepsilon_0\hbar c^3\}\left|\mu_{fi}\right|^2. \tag{12.26}$$

When we speak of the strength, or intensity of a particular transition, it is useful to introduce the dimensionless oscillator strength, f, of a transition:

$$f = \{4\pi m_e \nu_{fi}/3e^2\hbar\}\left|\mu_{fi}\right|^2. \tag{12.27}$$

The oscillator strength was derived classically by considering the time-dependent electric field of the incident radiation, $\mathcal{E}(t)$, driving or forcing the harmonic oscillations of a bound electron in the absorbing molecule. The relation between the oscillator strength and the integrated absorption coefficient (A) is $f = \{4m_e c\varepsilon_0/N_A e^2\}A$; the practical form of this expression is $f = 6.257 \times 10^{-19} A/(\text{m}^2\ \text{mol}^{-1}\ \text{s}^{-1})$ [4].

For a one-dimensional harmonic oscillator, $f = ⅓$ [4]. For an electron bound so that it oscillates harmonically in three dimensions, $f = 1$. The observed oscillator strength is therefore the ratio of the intensity of the transition to the intensity of a harmonically oscillating electron (in three dimensions). In practice, $f \approx 1$ for allowed electric dipole transitions and $f << 1$ for dipole forbidden transitions. It is worth comparing the above expressions for absorption and emission with the Planck law (see section 10.1), which describes the equilibrium between matter and energy. Spectroscopy derives from the energy density of radiating states, as defined in the explanation of black-body radiation; that is, from, $\rho_{\text{rad}}(E)$[1].

Given that the oscillator strength is something that can be measured, and is a dimensionless quantity, it is of interest to see if it is possible to obtain an expression

[1] It is important to recognise that different relations for the A and B coefficients can be found if other definitions of the radiation energy density are used. For example, B may be defined in terms of radiation density per unit wavenumber interval (a wavenumber is cm^{-1}, and so dimensionally, $\mathsf{L^{-1}}$), or radiation density per unit frequency interval. Then B could be derived in terms of irradiance (power per unit area per unit wavenumber) incident on the atom, or energy per unit volume per unit frequency interval on the atom. In addition, a factor of 4π (invisible to dimensional analysis) will appear in the definition of B if it is defined in terms of specific intensity (power per area per solid angle per frequency interval). The lesson to be learned here is that before one can make use of the various formulae for A and B coefficients from a particular source, one must carefully determine which measure of radiation intensity has been used. All too often, the same word 'intensity' is used to signify several quite distinct physical quantities—see table 12.2.

such as equation (12.27) by dimensional analysis using the Buckingham Π-theorem. The quantities that appear in equation (12.27) are described in table 12.3.

First, we decide which quantities will be important in defining the oscillator strength:

$$f \propto (m_e)^a(\nu_{fi})^b(e)^c(h)^d(\text{electric dipole})^e. \tag{12.28}$$

There are six variables, containing four base dimensions. Thus, from the Buckingham Π-theorem, there are 6 – 4 = 2 nondimensional Π-groups to be identified. First, let us consider Π_1. We will take the following variables as the repeat variables: m_e, ν_{fi}, e and h. Thus,

$$\Pi_1 = f(\mathsf{M})^a(\mathsf{T}^{-1})^b(\mathsf{TI})^c(\mathsf{ML}^2\mathsf{T}^{-1})^d.$$

But we see that f is dimensionless, and so $\Pi_1 = f$. We proceed in the same way to discover the second nondimensional group. This time, replacing the oscillator strength with the dimensions of an electric dipole.

$$\Pi_2 = (\mathsf{M})^a(\mathsf{T}^{-1})^b(\mathsf{TI})^c(\mathsf{ML}^2\mathsf{T}^{-1})^d\mathsf{LTI} = \mathsf{M}^{a+d}\mathsf{L}^{2d+1}\mathsf{T}^{-b+c-d+1}\mathsf{I}^{c+1}. \tag{12.29}$$

The values of the exponents are: $a = ½$, $b = ½$, $c = -1$ and $d = -½$, and we write for the second nondimensional Π-group, $\Pi_2 = \mu\sqrt{m_e}.\nu/e\sqrt{h} = \mu^2 m_e\ \nu/e^2h$. The Π-theorem allows us to manipulate the nondimensional groups as, $\Pi_1 = F(\Pi_2)$, where F is an unspecified function, and we may write for the oscillator strength of a transition,

$$f = \mu^2 m_e \nu / e^2 h. \tag{12.30}$$

This is the same as equation (12.27), but with all the numerical factors stripped away. Thus, demonstrating the simplicity of dimensional analysis, but also the disadvantage of dimensional analysis—a great deal of physics is lost; for example,

Table 12.3. Variables for the analysis of the strength of a transition.

Quantity/variable	Dimension	Description
$\lvert\mu_{fi}\rvert^2$	$(\mathsf{LIT})^2 = \mathsf{L}^2\mathsf{I}^2\mathsf{T}^2$	Square of the absolute value of the transition dipole moment of the transition of interest
m_e	M	Mass of the electron
ν_{fi}	T^{-1}	Frequency of the transition of interest
e	TI	Charge on the electron
h	$\mathsf{ML}^2\mathsf{T}^{-1}$	Planck constant
f		Oscillator strength
A	T^{-1}	Einstein A coefficient = $\{8\pi^2 v_{fi}^3/3\varepsilon_0\hbar c^3\}\lvert\mu_{fi}\rvert^2$
B	$\mathsf{M}^{-1}\ \mathsf{L}$	Einstein B coefficient = $(6\varepsilon_0\hbar^2)^{-1}\lvert\mu_{fi}\rvert^2$
ε_0	$\mathsf{M}^{-1}\mathsf{L}^{-3}\ \mathsf{T}^4\ \mathsf{I}^2$	Permittivity of free space

the transition dipole moment is a dipole moment arising from the mixing of the wavefunctions of two distinct quantum states, whereas μ^2 in equation (12.30) could be the square of the permanent dipole moment of the ground state. The two variables $|\mu_{fi}|^2$ and μ^2 have the same dimensions—but very different physics. Indeed, one often finds the term for the dipole matrix element, $|\mu_{fi}|^2$ written as $e^2. |<\psi_i| \boldsymbol{x} |\psi_f>|^2$. Here the dipole has been split into its component parts: a charge multiplied by a distance, or [electric dipole] = TI L, and the charge is now represented by the elementary charge, and the distance component by a matrix element coupling the amplitude of oscillation between an initial ψ_i and a final ψ_f state.

12.3.1 The effect of an electric field-gradient on allowed rotational spectra

As another application of the Buckingham Π-theorem to spectroscopy, we will generate an expression for the effect of an applied, static electric field-gradient on the rotational spectrum of a rigid-rotor. The effect of an electric field on allowed spectra was first described by Johannes Stark, and such investigations have yielded precision values of permanent molecular dipole moments. This problem, which the author worked on long-ago, involved calculating the effect of an inhomogeneous electric field on an allowed rotational (microwave) spectrum. The idea behind the project was to see if the application of a large, laboratory, electric field-gradient would perturb the energy states (ground and excited states) of the molecule under study sufficiently to modulate the intensity of the absorption. The applied field-gradient would interact with the permanent electric quadrupole moment of the molecule (if the molecule had a permanent dipole moment, it would also have a permanent quadrupole; e.g. HF, or CO; this not being the case in centro-symmetric molecules). Another aim of the project was to see whether this interaction (a lowering of the energy states involved in the transition, and a mixing of those states in the presence of the applied field-gradient) would modulate the observed intensity sufficiently to allow an additional route to the measurement of the molecular quadrupole moments [5].

We hypothesise that the applied electric field-gradient will modulate the amount of radiation absorbed by the sample; that is, the power absorbed. The rotational transition is allowed, so the oscillator strength (f) is of order unity in the absence of the applied field-gradient; however, it will be perturbed by the applied field-gradient. We hypothesise that in the presence of the applied field-gradient, the power absorbed (P) by the molecular sample will be (the details of the variables are given in table 12.4):

$P \propto (f)^a$ (molecular quadrupole moment)^b (applied electric field-gradient, E')^c $(c)^d$ $(\nu)^e$ (rotational constant, B)^f; that is: $P = \text{constant}\{(f)^a (\Theta)^b (E')^c (c)^d (\nu)^e (B)^f\}$.

In table 12.4 there are seven variables, containing four base dimensions. Thus, from the Buckingham Π-theorem there are three nondimensional Π-groups to be identified. First, let us consider Π_1; as the oscillator strength is without dimension we take $\Pi_1 = f$. We will take the following variables as the repeat variables: Θ, E', c and ν. Thus, the second Π-group is:

$$\Pi_2 = (\mathsf{L}^2\mathsf{TI})^b(\mathsf{MT}^{-3}\mathsf{I}^{-1})^c(\mathsf{LT}^{-1})^d(\mathsf{T}^{-1})^e\mathsf{ML}^2\mathsf{T}^{-3} = \mathsf{M}^{c+1}\mathsf{L}^{2b+d+2}\mathsf{T}^{b-3c-d-e-3}\mathsf{I}^{b-c}. \quad (12.31)$$

Table 12.4. Variables for investigating the perturbation of an allowed rotational transition.

Variable	Dimension	Description
P	$\mathsf{ML^2T^{-3}}$	Power of radiation absorbed (radiant flux, see table 12.2)
f		Oscillator strength of transition (related to the measured absorption)
B	$\mathsf{L^{-1}}$	Rotational constant
c	$\mathsf{LT^{-1}}$	Speed of light
ν	$\mathsf{T^{-1}}$	Frequency of the transition (in the microwave)
Θ	$\mathsf{L^2TI}$	Molecular quadrupole moment
E'	$\mathsf{MT^{-3}I^{-1}}$	Electric field-gradient

The solutions are: $b = c = -1$, $d = 0$ and $e = -1$; thus, we may write for the second nondimensional group: $\Pi_2 = P/\Theta E'\nu$. Dimensionally, this is power divided by energy ($[\Theta E'] = \mathsf{ML^2T^{-2}}$) multiplied by frequency. This procedure is repeated to find the third nondimensional group:

$$\Pi_3 = (\mathsf{L^2TI})^b(\mathsf{MT^{-3}I^{-1}})^c(\mathsf{LT^{-1}})^d(\mathsf{T^{-1}})^e\mathsf{L^{-1}} = \mathsf{M}^c\mathsf{L}^{2b+d-1}\mathsf{T}^{b-3c-d-e}\mathsf{I}^{b-c}. \tag{12.32}$$

This time, the various exponents are: $b = c = 0$, $d = 1$ and $e = -1$. Thus, the third nondimensional group is $\Pi_3 = Bc/\nu$.

Then, using the Π-theorem and the properties of the Π-groups, in particular, that we may invert one of the nondimensional groups (see chapter 5); $\Pi_2 = F(\Pi_1, \Pi_3) = F_1(\Pi_1, 1/\Pi_3)$. And the power absorbed by the molecular sample, in the applied electric field-gradient is:

$$P/\Theta E'\nu = F(\nu/Bc, f) \to P/\Theta E'\nu = (\nu/Bc)F_1(f) \to P = \{\Theta E'\nu^2/Bc\}F_1(f), \tag{12.33}$$

where we have an expression relating the intensity of the absorption of an allowed rotational transition to the oscillator strength of the transition, in the presence of an applied field-gradient. As for the full effect of an inhomogeneous electric field on the intensity of an allowed rotational transition; a great deal of algebra showed that for the P-branch, the total induced intensity (allowed by the electric dipole moment, μ, and induced by the coupling of the applied field-gradient and the molecular quadrupole moment, Θ) is given by [5]:

$$\begin{aligned} I_{\mathrm{P,z}} &= \frac{\Theta E'}{20B}\mu^2(\{J(J+1)/(2J+1)^2\} - \{J(J-1)/(2J-1)\}) \\ &= \frac{\Theta E'}{10B}\mu^2\{J/(2J+1)^2(2J-1)^2\}, \end{aligned}$$

where J is the rotational quantum number. The ratio of the induced to allowed intensity ($I_0 = ⅓\,\mu^2 J$) is [5]:

$$(I_{P,z}/I_{\mathrm{P},0}) = \frac{3\Theta E'}{10B}\left\{1/(2J+1)^2(2J-1)^2\right\}. \tag{12.34}$$

It can clearly be seen that the dimensional analysis, which gives the result $I \propto \Theta E'/Bc$, has lost all the numerical factors, but preserves the essential details.

12.4 Melting in organic solids

Dimensional analysis is used to investigate difficult, complex problems that are not readily susceptible to experimental investigation; particularly in new areas of research. This section will be an examination of such a use—we will look at the melting transition. The systems examined are a class of organic materials; consequently, the intermolecular bonding in the lattice consists of weak van der Waals forces. The solids consist of parallel, close-packed columns of alternating hexagonal molecules; the prototype of this class of layered, molecular binary-adducts is formed by mixing equimolar quantities of benzene and hexafluorobenzene (the structures of the four phases of the adduct given in figure 12.2) [6–8]. There are many such layered molecular adducts, and they nearly all exhibit solid-state phase transitions below their melting points—and they nearly all melt between about 273 and 325 K.

In these solids, the intracolumnar interactions are attractive electrostatic forces developed between large alternating positive (hexafluorobenzene) and negative (benzene) molecular quadrupole moments [8]. Intercolumnar interactions arise between the bond dipole moments of the $^{(\delta+)}$C-F$^{(\delta-)}$ bonds in one column and the adjacent $^{(\delta+)}$H-C$^{(\delta-)}$ bond dipoles in neighbouring columns, and give lateral stability [8]. Such intercolumnar interactions are simple electrostatic, dipole–dipole interactions varying as $1/r^3$; where r is the intermolecular separation. These intercolumnar interactions are clearly seen in the lowest temperature phase (IV), see figure 12.2 (lowest image). The columns of hexagonal molecules are like interlocking cogs and wheels in a clockwork mechanism. But as the temperature increases the columns separate, as the weak attractive intermolecular forces are overcome by the available thermal energy, and the cogs are able to move independently (a rotational relaxation) on their columns. A solid-state phase transition is this concerted motion taking place; when the thermally-driven mean amplitudes of oscillation reach a certain magnitude (see figure 12.3). And the solid melts at the temperature when the available thermal energy exceeds the total attractive intermolecular forces.

One of the earliest theories of melting is due to Lindemann in 1910 [9]. He suggested that for all solids, due to thermally-driven fluctuations of the position of atoms in a solid, melting occurred at a temperature (T_m) when the average displacement of the atom, $<u>$, which constitute the solid reaching a given fraction (say 10%, $r^* \approx 0.1$) of the average intermolecular or interatomic spacing, r; that is, $r^* = \sqrt{< u^2 >_{Tm}}/r \approx 0.1$[2].

The first dimensional analysis on what constituted melting was by Albert Einstein, who in 1911 [10] investigated three problems in the physics of solids: the characteristic frequency of the lattices of crystalline, atomic solids as a function of their moduli of compressibility, of their melting points, and the thermal conductivity of crystalline insulators. Recognizing that temperature does not have the same dimension as energy, and before the acceptance of the existence of molecules, Einstein had recourse to the artifice of replacing the dimension for temperature, θ, by

[2] This is what is termed a heuristic model. The original ideas of Niels Bohr about the structure of the hydrogen atom was another such heuristic model.

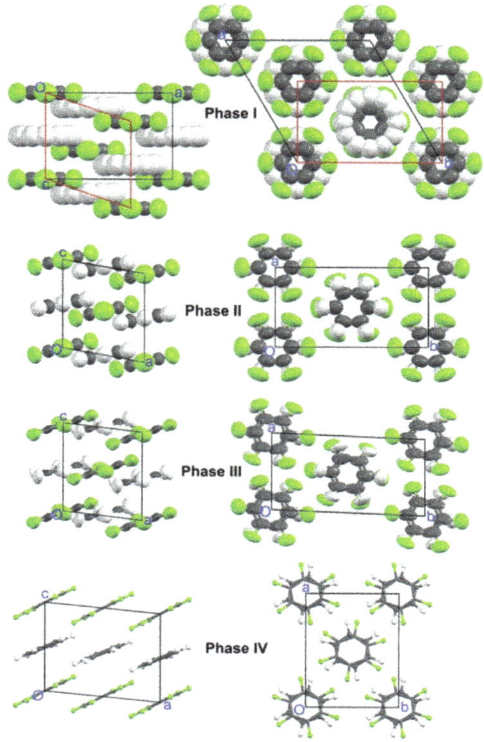

Figure 12.2. X-ray crystal structures of the four phases of C_6H_6:C_6F_6 [8]; phase IV is the lowest temperature phase. Interatomic distances are determined with a standard uncertainty of about 0.007 Å. Fluorine atoms are represented in green, hydrogen atoms are represented in white, and grey is used for the carbon atoms. The 'size' of the atom corresponds to levels of thermal excitation. The molecules are free rotors in the highest-temperature phase. Image reproduced from [8]. CC BY 3.0.

its product with the Boltzmann constant, k_B; thereby obtaining useful results. This difference between the dimensions of heat and energy, that is [Temperature] = θ and [Energy] = ML^2T^{-2}, is at the heart of the early use of dimensional analysis, and was the basis for the Rayleigh–Riabouchinsky Paradox (see section 7.4).

We wish to apply dimensional analysis to the process of melting—so we must first identify which physical quantities are important in the melting process. We will look at the organic solids mentioned above, and for which we have extensive experimental data on the temperature dependence of the location and dynamics of the molecules in the lattice (figures 12.2 and 12.3).

Let us first try a straightforward application of dimensional analysis (the Rayleigh algorithm) to the melting temperature, T_m, of an organic binary-adduct. We hypothesise that the magnitude of the melting temperature, which is a measure of the thermal energy in the solid required to overcome, or at least destabilise, the intermolecular forces holding the solid together, $E(r)$, is proportional to the magnitude of the average amplitude of motion of the molecules, $<u>$, as they librate, or oscillate on their lattice sites, the Planck constant, h, the mass of the

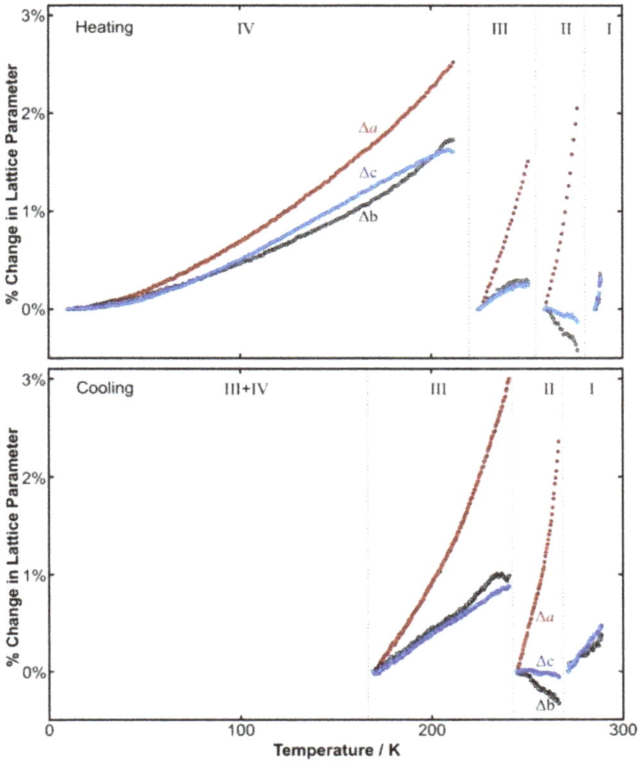

Figure 12.3. Percentage change of lattice parameters (***a***, ***b*** and ***c***) across the temperature range of C_6H_6:C_6F_6 (phases I to IV)—this is a measure of the temperature dependence of the mean amplitudes of oscillation of the molecules in the three crystallographic directions. Full details of how this data was measured and treated are to be found in [8]. (There is a significant hysteresis in the dynamics of the phase transitions of this solid.)

liberating molecule, m, and the Boltzmann constant, k_B. Thus, we hypothesise that $T_m \propto (u)^a(h)^b(m)^c(k_B)^d$. The dimensional equation is then written as (the dimensions for this problem are to be found in table 12.5):

$$\theta = (\mathsf{L})^a(\mathsf{ML^2T^{-1}})^b(\mathsf{M})^c(\mathsf{ML^2T^{-2}\theta^{-1}})^d = \mathsf{M}^{b+c+d}\mathsf{L}^{a+2b+2c}\mathsf{T}^{-b-2d}\theta^{-d}. \qquad (12.35)$$

The solutions are: $a = -2$, $b = 2$, $c = -1$ and $d = -1$. We may then write an expression for the melting temperature of the solid in terms of the above four variables as:

$$T_m = h^2/ < u^2 > mk_B, \text{ or } < u^2 > = h^2/mk_BT_m. \qquad (12.36)$$

This simple application of dimensional analysis has generated an expression that may also be derived from the Debye theory of heat capacities. If we assume a Debye solid and neglect the frequency difference between longitudinal and transverse phonons, we may write [11]:

Table 12.5. Variables for the thermal dynamics of solids.

Variable	Dimension	Description
$E(r)$	ML^2T^{-2}	Intermolecular interaction energy (potential)
m	M	Mass of molecules composing the solid
L	L	Spacing between molecules in solid
k_B	$ML^2T^{-2}\theta^{-1}$	Boltzmann constant
ν	T^{-1}	Frequency of low-frequency libration
$<u>$	L	Mean amplitude of oscillation of low-frequency libration
T_m	θ	Melting temperature
h	ML^2T^{-1}	Plank constant
$<\varphi>$		Mean amplitude of angular oscillation of a molecule in the solid (like a rotation)

$$<u^2>_{Tm} = 9h^2T_m/4\pi^2mk_BT_D^2. \tag{12.37}$$

Here, T_D is the Debye temperature, and m is the molecular mass (the Debye temperature is approximately equal to the phonon energy of the minimum wavelength mode, and so we can interpret the Debye temperature as the temperature at which the highest frequency (lowest wavelength) mode, and hence every other mode is excited, $T_D = \hbar\omega_D/k_B$); i.e., the highest temperature that can be achieved due to a single normal vibration. Stripping away the numerical factors, which are invisible to dimensional analysis, and noting that dimensional analysis will only give you the lowest-order combination of variables needed to generate a particular desired combination, the dimensional analysis has taken the two temperatures in equation (12.37) and assumed they are identical (they have the same dimension) to give $T_m/T_D^2 \sim 1/T$. Dimensional analysis gives only the bare bones of an expression, and so should be used with caution. But with physical insight, we can use dimensional analysis to analyse data.

Consider a further application of dimensional analysis; this time looking for the amplitudes of the librations, or excitations perpendicular to the six-fold rotational axis of the benzene and hexafluorobenzene molecules in the solid, $<\varphi>$ (the temperature dependence of such data, measured by Raman spectroscopy [12], is given in figure 12.4). The maximum magnitude of this amplitude is [11], of order, 7.9° (given the geometry of the molecules, this amplitude of oscillation is a significant fraction of the equivalent 60° positions). This maximum of librational amplitude is observed just below the melting point of the solid [11].

Again, we hypothesise the origin of this rotational amplitude (the dimensions are given in table 12.5),

$$<\varphi> \propto (m)^a(k_B)^b(L)^c(\upsilon)^d(\text{Temperature})^e. \tag{12.38}$$

There are six variables, made up of four base dimensions, so we will use the Buckingham Π-theorem to derive an expression relating amplitude to temperature

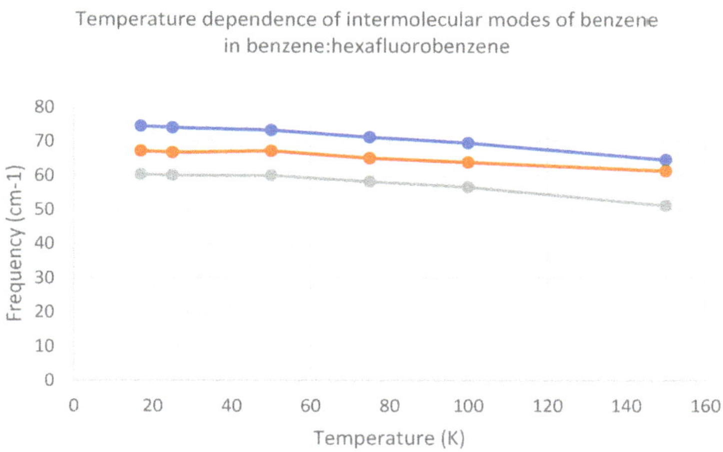

Figure 12.4. The temperature dependence of the frequencies of the librational transitions of the benzene molecule in phase IV of the binary-adduct, C_6H_6:C_6F_6. The spectra were recorded with a high-resolution laser Raman spectrometer [12]. The two highest frequency lines (blue and orange data; are the two perpendicular excitations; the grey data is the parallel excitation) are seen to become degenerate as the sample temperature increases. In fact, they become equal at the transition between phase IV and phase III [12].

based on two nondimensional Π-groups. In this analysis we will use: (m), (k_B), (L) and (υ) as repeat variables. As the rotational amplitude is without dimension, we take $\Pi_1 = \langle\varphi\rangle$.

The second nondimensional group is

$$\Pi_2 = (\mathsf{M})^a(\mathsf{ML^2T^{-2}\theta^{-1}})^b(\mathsf{L})^c(\mathsf{T}^{-1})^d\theta = \mathsf{M}^{a+b}\mathsf{L}^{2b+c}\mathsf{T}^{-2b-d}\theta^{-b+1}.$$

The solutions are: $a = -1$, $b = 1$, $c = -2$ and $d = -2$, and the second nondimensional Π-group is

$$\Pi_2 = k_BT/mL^2\upsilon^2.$$

Dimensionally, this expression is thermal energy/rotation-like energy, and so is dimensionless. The Buckingham Π-theorem allows us to write $\Pi_1 = f(\Pi_2)$, and so,

$$<\varphi> = k_BT/mL^2\upsilon^2. \tag{12.39}$$

The amplitude of the rotation-like libration is given by the available thermal energy, weighted by the mechanical (kinetic) dynamics of the molecule in the lattice. The lower the frequency of the librational excitation, the larger the amplitude, and the more it will contribute to the melting temperature. It should be noted that the term mL^2 has the same dimensions as the moment of inertia of the liberating molecule (I); that is, ML^2, or kg m^2. Thus, $\langle\varphi\rangle = k_BT/I\upsilon^2$.

Using dimensional analysis, we have derived two simple expressions telling us how the amplitudes of motion of the molecules in the solid (as measured in

Table 12.6. The numerical factors lost by dimensional analysis.

	Expression derived by dimensional analysis	Expression derived by the full theory
Mean amplitude of oscillation	$<u^2> = h^2/m\, k_B T_m$	$< u^2 >_{Tm} = 9h^2 T_m / 4\pi^2 m k_B T_D^2$
Mean amplitude of libration (a rotation-like oscillation). The method of estimating $<\varphi>$ by dimensional analysis cannot distinguish between $<\varphi>$ and $<\varphi^2>$	$<\varphi> = k_B T / I\nu^2$	$<\varphi^2> = k_B T / 4\pi^2 I\nu^2$

figure 12.3) scale as the sample temperature changes, and how the amplitude of a rotation-like motion scales with temperature, and the frequency of the librational modes:

$$T_m = h^2 / < u^2 > mk_B, \text{ or } < u^2 > = h^2/mk_B T_m; \text{ and} \quad (12.40)$$

$$<\varphi> = k_B T / I\nu^2, \text{ thus, } k_B T = <\varphi> I\nu^2. \quad (12.41)$$

Table 12.6 shows the numerical factors that have been lost by dimensional analysis.

We will now hypothesise that the frequency of a lattice excitation (ν) in a solid binary-adduct is proportional to the potential energy between the atoms or molecules, $E(r)$, comprising the lattice, the mass of the molecules (m), and the distance (L) between those constituent atoms or molecules; that is,

$$\nu \propto (E(r))^a (m)^b (L)^c. \quad (12.42)$$

This leads to a dimensional equation (the dimensions are given in table 12.5),

$$\mathsf{T}^{-1} = (\mathsf{ML^2T^{-2}})^a (\mathsf{M})^b (\mathsf{L})^c = \mathsf{M}^{a+b}\mathsf{L}^{2a+c}\mathsf{T}^{-2a}.$$

The solutions are: $a = ½$, $b = -½$ and $c = -1$. We find, therefore, that based on our hypothesis the frequency of a lattice mode is given by

$$\nu = \frac{1}{L}\sqrt{\left(\frac{E(r)}{m}\right)}, \text{ or } \nu^2 = \mathrm{E(r)} / L^2 m. \quad (12.43)$$

Given that $L^2 m$ has the same dimensions as the moment of inertia, I, (kg m^2), we may write $\nu^2 = E(r)/I$. In these solids, we know that the attractive intercolumnar interaction and the intracolumnar interaction are determined by the bond dipole moments (μ_1 and μ_2) and the permanent molecular quadrupole moments (Θ_1 and Θ_2) [6]. Thus, inserting our value for the intermolecular potential for the binary-adducts of interest, which we have identified by dimensional analysis (section 9.3), for example, $E(r) = \mu_1\mu_2/\varepsilon_0 r^3 + 2\mu\Theta/\varepsilon_0 r^4 + \Theta_1\Theta_2/\varepsilon_0 r^5$, we have an expression for the frequency of a lattice vibration of a solid formed by the electrostatic interactions between dipole moments and quadrupole moments: $\nu = \frac{1}{L}\sqrt{E(r)/m}$. This expression could now be tested via experiment and detailed theoretical calculations. It is a first step in relating the frequencies of lattice vibrations to the temperature of the lattice; that is, to the energy content of the lattice, and to the presence or absence of phase transitions and to the melting point of the lattice.

From a harmonic oscillator, we know $\upsilon^2 = E(r)/I$; that is, the frequency of the intermolecular mode is determined by the magnitude of the intermolecular potential within which the liberating molecule finds itself, $E(r)$. The magnitude of this potential will diminish (the energy surface will become flatter), relative to the available thermal excitation as the temperature of the sample rises. As the temperature mounts towards the melting temperature, $\nu \rightarrow 0$. This is what is observed in the data in figure 12.4. Similarly, it is possible to observe rapid variations in the magnitudes of the mean squared amplitudes of the molecules below the temperature of phase transitions, see figure 12.3.

The crystal structure [8] reveals that the environment of the molecules in phase IV is anisotropic (orthorhombic, $P2_1/a$); that is, the intermolecular potential is also anisotropic, and there will be three different force constants and three different librational frequencies for each of the molecules, of the same chemical type, in the unit cell in phase IV; one parallel vibration and two perpendicular vibrations. The axis of the vibrations is the six-fold rotation axis of the stacked molecules in the column, and is parallel to the c-axis of the lattice [8]. As the temperature rises, there are phase transitions which break the weak, electrostatic intercolumnar interactions; thereby facilitating the rotation-like oscillations of the benzene and the hexafluorobenzene molecules. In the narrow highest-temperature phase (I), the alternating, equidistant benzene and hexafluorobenzene molecules are free rotors [8].

Using the full expression for $<\varphi^2>$ in table 12.2, we may write for the temperature of this maximum amplitude of oscillation, T_m,

$$T_m = {\varphi_{\max}}^2 4\pi^2 I\upsilon^2/k_{\mathrm{B}}, \tag{12.44}$$

the melting temperature. It is seen [11] that using the known melting point of solid benzene and C_6H_6:C_6F_6 it is possible to calculate frequencies of librations that agree with those measured experimentally [11]. In any investigation of the origin of $<u>$ and $<\varphi>$, it is the lowest frequency, intermolecular vibration that will dominate. Consequently, in searching for an explanation of the melting temperature of a solid, it is the lowest frequency vibrations that need to be considered. This weighting in favour of low frequency, and hence large mean squared vibrational amplitudes reflects the coupling of the motions of neighbouring molecules.

This simple model of melting tells us that melting of the solid composed of small aromatic molecules occurs when the constituent molecule is no longer able to store the increasing amounts of available thermal energy in the manifold of excitations associated with the low-frequency librations, or rotations of the molecules within their lattice positions. As a consequence, this energy excites the vibrations of the weak intermolecular bonds, thus destabilising the lattice. The available thermal energy is rapidly transferred from the vibrating, or rocking molecules to the weak intermolecular bonds, leading to the break-up of the intermolecular bonding and nence to melting. The data suggest that a model like the above model for the melting of solids composed of small aromatic molecules in terms of the mean amplitudes of libration or oscillation may be constructed for each of the solid-state phases for such solids. That is, when the amplitude of oscillations of entire molecules in fixed lattice sites reaches a certain maximum (see figure 12.3), that particular structure, or polymorph becomes unstable, relative to a new crystal structure at that temperature, and a phase transition occurs.

References

[1] Rutherford E 1911 The scattering of α and β particles by matter and the structure of the atom *Phil. Mag.* **21** 669

[2] Bohr N 1913 On the constitution of atoms and molecules *Phil. Mag. Series* **6** pp 1–25

[3] De Broglie L 1925 Recherches sur la théorie des quanta *Ann. Phys.* **3** 22–128

[4] *Molecular Quantum Mechanics* ed Atkins P and Friedman R 2005 (Oxford: Oxford University Press)

[5] Williams J H 1992 The effect of an electric field gradient on allowed rotational spectra *J. Phys. B: At. Mol. Opt. Phys.* **25** 145–54

[6] Williams J H 1993 The molecular electric quadrupole moment and solid-state architecture *Acc. Chem. Res.* **26** 593–98

[7] Williams J H, Cockcroft J K and Fitch A N 1992 Structure of the lowest temperature phase of the solid benzene–hexafluorobenzene adduct *Angew. Chem. Int. Ed. Engl.* **31** 1655–657

[8] Cockcroft J K, Rosu-Finsen A, Fitch A N and Williams J H 2018 The temperature dependence of C–H···F–C interactions in benzene:hexafluorobenzene *CrystEngComm* **20** 6677–682

[9] Lindemann A F 1910 The calculation of molecular vibration frequencies *Phys. Z.* **11** 609–12 Lawson A C 2009 Physics of the Lindemann melting rule *Philos. Mag.* **89** 1757–70

[10] Einstein A 1911 Elementare Betrachtungen über die thermische Molekularbewegung in festen Korpern (Elementary observations concerning the thermal molecular motion in solid bodies) *Ann. Phys.* **35** 670–94

[11] Williams J H and Frick B 1992 Low frequency excitations of aromatic molecules in the solid state *Chem. Phys.* **166** 425–39

[12] Williams J H and Becucci M 1993 Vibrational Raman spectroscopy of solid benzene: hexafluorobenzene *Chem. Phys.* **177** 191–202

IOP Publishing

Dimensional Analysis
The great principle of similitude
Jeffrey H Williams

Chapter 13

Modelling phenomena

'Do not imagine that mathematics is harsh and crabbed, and repulsive to common sense. It is merely the etherealisation of common sense'.
Lord Kelvin, From an address in Birmingham (1883)

In theory, science progresses by a simple process. We construct a hypothesis, which we test by experiment. If the result of our experiment validates our initial hypothesis—the problem is solved. If the experiment does not validate our initial hypothesis—we need to think about our initial hypothesis more deeply. This is the manner in which science has advanced, since the ancient Greek natural philosophers. In this chapter we look at prototypes. These are complex experiments—multidimensional experiments, so they do not give a simple, single-valued answer to a hypothetical question. It may even be the case that a series of experiments, or rather, the results from a series of experiments may be needed to better formulate the initial hypothesis. As Orville Wright put it, '*At first, we had taken up the problem* [of heavier than air flight] *merely as a matter of sport, but now it became apparent that if we were to make much progress it would be necessary to get better tables from which to make our calculations. In September we set up a small wind-tunnel… We made thousands of measurements of the lift, and of the ratio of the lift to the drift.*' [1]

13.1 Prototypes

A prototype is an early, full-scale working model of a machine, or a system; for example, a prototype could be a full-scale functioning airplane, or merely a scaled model of a new type of submarine or sailing boat. We define the prototype to be a model, in that it refers to a construct used to predict the behaviour, or performance of the prototype; thereby saving time and resources in the development of the final product. But its form and weight will be a scaled-down representation of the final

doi:10.1088/978-0-7503-3655-0ch13

product. Today, if we were following the Wright brothers in developing a new airplane, we would fabricate a model of the airplane in polymer, on a 3D-printer, and then place this model in a small wind-tunnel.

The rules relating the behaviour of models and prototypes to the final product are termed similarity rules. The purpose of these rules is to quantify the relationship, via measurements of the models, to predict the corresponding behaviour of the prototype. These experimentally defined relationships are termed scaling laws between model and prototype. The question addressed by such scaling laws is: What aspects of the model and the prototype should be similar, permitting us to make predictions about the prototype—based on an experimental study of the model? There are three similarity conditions that must be met:

1. Geometric similarity: This is the obvious requirement that the model and the prototype have the same form. So, for example, the ratios of lengths of respective features and all of the angles are the same in both model and prototype. Such a similarity of structure will ensure that the velocity and dynamics of a flow of the medium in which the model and the prototype are to be placed will be the same, or can, at least, be related.

   Idealized spheres are always geometrically similar, while humans are only approximately geometrically similar. A scale model of the Boeing 747 with a 20 cm wing-span is made to be geometrically similar to the actual 64 m wing-span of the full-scale aircraft [1]. Once we have applied a geometrical similarity, we are able to describe the geometry by a single length-scale; for example, wing-span or axial length.
2. Kinematic similarity: This requires the velocities of fluids (vector quantities with magnitude and direction) interacting with homologous points on the surface of the model and the prototype be in the same direction, and have magnitudes that are related by scaling factors. The requirement of kinematic similarity is the reason for geometric similarity; for example, in fluid flow the body's surface generates the distribution of streamlines. Kinematic similarity requires that the flow regimen be relatable between model and prototype; for example, if compressibility, surface tension, or free energy effects are calculated to be absent from the flow around the prototype, they must also be absent in the flow patterns for the experiments on models derived from the prototype.
3. Dynamic similarity: This requires that both model and prototype experience identical forces, at corresponding points, and that these forces be parallel, and be related by a constant scaling factor throughout any flow fields. This is the most restrictive requirement, since dynamic similarity requires both geometric and kinematic similarity. If these three rules of similarity are followed, the model and its dynamic behaviour will be related directly to the same properties of the prototype.

13.2 Experimental design and interpretation

Let us look briefly at deriving similarity laws. Consistent with the use of the Buckingham Π-theorem, we will use the term, Π_i, to indicate the ith nondimensional group. For dynamic similarity, each value of a nondimensional parameter of the model must match the respective value for that parameter evaluated for the prototype. For m Π-groups, associated with the underlying physics of the problem under investigation, we have:

$$\Pi_{1,\text{ model}} = \Pi_{1,\text{ prototype}}$$

$$\Pi_{2,\text{ model}} = \Pi_{2,\text{ prototype}}$$

$$\dots$$

$$\Pi_{m,\text{model}} = \Pi_{m,\text{prototype}}. \tag{13.1}$$

Thus, if we were analysing the drag force on a new design of submarine, the above mapping of the nondimensional groups would ensure that:

$$\text{Re}_{\text{model}} = \text{Re}_{\text{prototype}}, \text{ and } C_{\text{d,model}} = C_{\text{d,prototype}}; \tag{13.2}$$

for Reynolds number and drag coefficient, C_{d}. The hypothesised functional relationship implies that we must match all but one nondimensional parameters. Consider the functional relationship for nondimensional groups of variables in either prototype or model:

$$\Pi_m = F(\Pi_1, \Pi_2, \dots \Pi_{m-1}). \tag{13.3}$$

If our hypothesis is correct, then knowledge of $\Pi_1, \Pi_2, \dots \Pi_{m-1}$ determines Π_m. Hence, we can say that if $F(\Pi_1, \Pi_2, \dots \Pi_{m-1})$ for the model is known, then Π_m for the model is also known. If we match $(\Pi_1, \Pi_2, \dots \Pi_{m-1})$ for the model to the prototype, then Π_m will also be matched, since

$$\Pi_{m,\text{prototype}} = F(\Pi_{1,\text{prototype}}, \Pi_{2,\text{prototype}}, \dots \Pi_{m-1,\text{prototype}}). \tag{13.4}$$

We can define similarity, or similitude, as the condition that each of the nondimensional groups of variables defining the behaviour of the model be the same as each respective nondimensional parameter for the prototype. So, even though we are undertaking experiments on models, the results are applicable to the prototype—which will lead us to the final product.

13.2.1 Scaling the prototype of a submarine

Let us assume that we are designing a submarine. The essential feature of all submarines is that they are long and oval in shape——thereby minimizing the drag force—and that the geometry of the submarine can be characterized by a single length parameter, l. We then postulate that the key variables in determining the drag force, F_{d}, of all submarines of this shape are: length-scale, l, velocity, v, fluid density, ρ and fluid dynamic viscosity, μ. We wish to know the drag force, and we begin by hypothesizing that the drag of this shape of submarine when submerged will likely be of the form:

$$F_{\text{d}} = f(l, v, \rho, \mu). \tag{13.5}$$

Having fixed the shape of the submarine, the single length-scale (l) completely describes every aspect of its geometry.

We wish to explore a multidimensional problem; that is, to measure the drag force as a function of the variables given in equation (13.5). So, we would need experimental studies of F_d versus l, versus v, versus ρ, and versus μ in all possible combinations. Alternatively, we can use dimensional analysis and construct a model. Dimensional analysis will quickly tell us if our initial hypothesis about the best design for a submarine is correct or not. And a substantial programme of experimentation could be avoided, as we can quickly show the relationship between those variables. We see in equation (13.5) that there are five variables, involving the three mechanical dimensions. It is then straightforward to find the two nondimensional Π-groups involved in this study. The dimensions of the variables are given in table 13.1.

Using the techniques of dimensional analysis outlined in this volume, the two groups of variables, which generate dimensionless variables are: $\rho v^2 l^2$ with dimensions of a force, and ρvl, which has the same dimensions as μ. Thus, we may write

$$F_d/\rho v^2 l^2 = F\left(\frac{\rho v l}{\mu}\right). \tag{13.6}$$

So, we know how the variables are related. Some experiments are still required, however, and to undertake the experimental investigation of our model, we must:

- Chose a suitable fluid support; that is, a single ρ (for example, dry air at atmospheric pressure and ambient temperature: 1.2041 kg m^{-3} at 293 K and 1 atm.) and μ (dynamic viscosity of dry air = 1.825×10^{-5} kg m^{-1} s^{-1}, and kinematic viscosity of dry air = 1.516×10^{-5} m^2 s^{-1} at 293 K and 1 atm. The difference between the values of the dynamic and kinematic viscosity is ρ.)
- Build a small-scale submarine; perhaps 10 cm in length by, for example, 3D printing.
- Measure, of order, ten vales of v for ten values of F_d and plot the data in terms of the nondimensional variables associated with the function F in equation (13.6). Use this plot to estimate the function F.

Experiments will also reveal any inadequacies or inconsistencies in the initial hypothesis; for example, have we included a superfluous variable, and/or are we

Table 13.1. Variables for modelling a submarine (the variables in equation (13.5)).

Variable	Dimension	Description
F_d	MLT^{-2}	Drag force
v	LT^{-1}	Velocity of supporting medium
ρ	$M\ L^{-3}$	Density of supporting medium
μ	$ML^{-1}\ T^{-1}$	Viscosity of medium
l	L	Length-scale of submerged vessel

missing an essential variable? If we have a superfluous dependent variable in the function F, we will find that F has little or no dependence on this variable over the range of interest—the data will not match our theoretical model. For example, if we added the speed of sound, c, to our analysis of the submarine design, so the drag force would now be dependent upon

$$F_d = f(l, v, \rho, \mu, c). \tag{13.7}$$

Dimensional analysis would yield an additional dimensionless Π-group, compared to equation (13.6):

$$F_d/\rho v^2 l^2 = \mathrm{F}\left(\frac{\rho v l}{\mu}, \frac{v}{c}\right), \tag{13.8}$$

where v/c is known as the Mach number (Ma). This quantity is relevant only when we are considering supersonic speeds. In the design of a new submarine, where velocities will be, of the order, a few tens of metres per second when submerged, we would discover that the measured drag term $F_d/\rho v^2 l^2$ would be insensitive; that is, very weakly dependent on the value of the ratio v/c, if the ratio $\frac{\rho v l}{\mu}$ were held constant.

Let us suppose we have built a model of a submarine, a 1/10 scale model of the prototype, via 3D-printing. We wish to measure the properties of this model in an experiment. It would be cheaper and easier to use a wind-tunnel than a water-bath, so we need to consider how this constraint might impact on the measurements. To estimate the drag force versus the velocity curve for the prototype, which will 'travel' at $0.5\ \mathrm{m\ s^{-1}}$; at what air velocity must the wind-tunnel be run to mimic this velocity? How to relate the drag of the model to that of the prototype? Suppose the only information we have is given in table 13.2 (where l is the scaled dimension, v is the velocity of the supporting medium and υ is the kinematic viscosity defined as $\upsilon = \mu/\rho$).

As we have seen in equation (13.6), the relevant function relating the nondimensional groups in this problem is (remember, $\upsilon = \mu/\rho$),

$$F_d/\rho v^2 l^2 = \mathrm{F}\left(\frac{v l}{\upsilon}\right). \tag{13.9}$$

We have formatted the Reynolds number in terms of the kinematic viscosity. For dynamic similarity, we know that, if we match the Reynolds number the drag coefficient will also be matched. Hence, we rewrite the similarity laws as:

Table 13.2. Some properties of a model and a prototype at ambient temperatures (from [1]).

Fluid support of model		Fluid support of prototype
	$10 l_{\mathrm{model}} = l_{\mathrm{prototype}}$	
$\upsilon_{\mathrm{model}} = 1.516 \times 10^{-5}\ \mathrm{m^2/s}$ (in air)		$\upsilon_{\mathrm{prototype}} = 1.002 \times 10^{-6}\ \mathrm{m^2/s}$ (in water)
$v_{\mathrm{model}} = ?$		$v_{\mathrm{prototype}} = 0.5\ \mathrm{m\ s^{-1}}$

$$\text{if} \left.\frac{vl}{\upsilon}\right|_{\text{model (m)}} = \left.\frac{vl}{\upsilon}\right|_{\text{prototype (p)}}, \text{ then } F_d/\rho v^2 l^2\Big|_{\text{model (m)}} = F_d/\rho v^2 l^2\Big|_{\text{prototype (p)}} \cdot \quad (13.10)$$

Then we calculate, $v_m = v_p\ (l_p\ \upsilon_m/l_m\ \upsilon_p) = 75\ \text{m s}^{-1}$. That is, our model is inappropriate. A 1/10 scale model of the prototype, coupled with the kinematic viscosity of water generates a fluid forcefield that is so much smaller than that of air ($\upsilon_m/\upsilon_p = 14.95$) that we would be required to run the model in the wind-tunnel at a velocity 14.95 times higher than the prototype velocity in a water-bath. If the wind-tunnel cannot handle this high velocity, then we could investigate the model using a different support medium, or build a larger scale model. If the wind-tunnel is able to handle this velocity then we would be able to relate the drag from the similarity laws:

$$F_{d,p} = (\rho_p v_{p^2} l_{p^2}/\rho_m v_{m^2} l_{m^2}) F_{d,m}, \tag{13.11}$$

which is a factor 3.72 larger.

13.2.2 Modelling boats

When it comes to surface vessels, the situation is more complex. Hull drag in a surface boat is more difficult to model, because the drag force arising from a partially submerged hull moving through the water is a function of: the velocity of the boat relative to the water, the orientation of the hull (for example, is it a sailing boat heeling over, see figure 13.1), and the conditions of the surface water—are there big waves, or small waves? Any image of a sailing boat under sail will illustrate this complexity.

We will begin the dimensional analysis of the problem of a surface boat by considering an upright boat moving steadily on calm water. If we wish to design a new sail-boat, we must build a model to represent a prototype, and then devise an experimental situation where the dynamical properties of the model, and hence the prototype may be studied. Hull drag is due to both shear forces of viscous drag and to pressure forces on the hull. These pressure forces are due to the coupling of the velocity of the water and the deformable shape of the water against the near-rigid, solid outline of the hull. Even in the calmest conditions, the motion of the hull creates waves, and disturbs the surface of the water. The forward motion of the boat is being resisted by a wave of its own making. Table 13.3 contains the variables needed to investigate the motion of a surface boat.

We have the usual three mechanical base dimensions M, L and T, which form the six variables in the problem. So, by the Buckingham Π-theorem, we will have three dimensionless variables to identify. We choose v, l and ρ as our repeating variables, which involve all three dimensions. Dimensional analysis yields (as in the previous example in this chapter—the submerged submarine):

$$\Pi_1 = F_d/\rho v^2 l^2 \text{ and } \Pi_3 = \mu/\rho v l, \tag{13.12}$$

which give the dimensionless drag and dimensionless viscosity—the inverse of the ship's Reynolds number. This leaves us with the third Π-group to be found, which involves the acceleration due to gravity.

Figure 13.1. Wooden sailing boat Kleine Freiheit—70-year-old gaff cutter—operation: Abenraa, Denmark. This 'a gaff cutter' image has been obtained by the author from the Wikimedia website where it was made available under a CC BY-SA 3.0 license. It is included within this article on that basis. It is attributed to Uwe Kils.

Table 13.3. Variables for the motion of a sail-boat.

Variable	Dimension	Description
F_d	MLT^{-2}	Drag force on boat
v	LT^{-1}	Velocity of boat
l	L	Length of boat (water-line)
g	LT^{-2}	Acceleration due to gravity
ρ	ML^{-3}	Density of supporting fluid
μ	$ML^{-1}\ T^{-1}$	Viscosity of fluid

$\Pi_2 = (v)^a(l)^b(\rho)^c\ g$, so that the dimensional equation is

$$M^\circ L^\circ T^\circ = (LT^{-1})^a(L)^b(ML^{-3})^c\ LT^{-2} = M^c\ L^{a+b-3c+1}T^{-a-2}. \tag{13.13}$$

From this we have: $a = -2$, $c = 0$ and $b = 1$; giving $\Pi_2 = gl/v^2$, which is dimensionless gravity, or the Froude number (Fr) of the ship raised to the power -2 (see section 4.1). Using the properties of the dimensionless Π-groups, we may invert them to give the more familiar form—see chapter 5. Hence, the final result for this problem is:

$$F_d/\rho v^2 l^2 = f\left(\frac{\mu}{\rho v l},\ gl/v^2\right). \tag{13.14}$$

Multiplying the LHS by ½ and re-arranging allows us to write for the drag coefficient

$$C_d = G(Re,\ Fr^2), \tag{13.15}$$

where we see the trade-off between the kinetic and the potential energy (here of wave motion) as characterized by the Froude number, Fr. For dynamic similarity, we require the following similarity laws between model and prototype:

$$\left.\frac{\rho v l}{\mu}\right|_m = \left.\frac{\rho v l}{\mu}\right|_p, \text{ and } (v^2/lg)|_m = (v^2/lg)|_p. \tag{13.16}$$

Given our hypothesised function, we have:

$$F_d/\rho v^2 l^2\,|_m = F_d/\rho v^2 l^2\,|_p. \tag{13.17}$$

Consider a typical set of data [1]. A model boat of a scale 1:50 was measured in a towing tank to have a drag force of 0.27 N when towed at a velocity of 0.91m s^{-1}. If we assume that most of this drag is not due to friction, but to generated waves we can estimate the drag force that would arise on the prototype boat. If we ignore viscosity, then for dynamical similitude we must have both $F_d/\rho v^2 l^2$ and gl/v^2 to be the same in each case. Considering the latter, we have: $gl_m/v_m{}^2 = gl_p/v_p{}^2$, such that $(v_p/v_m) = \sqrt{g}\,(\sqrt{l_p/l_m}) = \sqrt{50}$. So, $v_p = \sqrt{50}\,v_m$ and for the force, $F_{d,\ m}/\rho v_m{}^2 l_m{}^2 = F_{d,\ p}/\rho v_p{}^2 l_p{}^2$; such that, $(F_{d,\ p}/F_{d,\ m}) = \rho v_p{}^2 l_p{}^2 / \rho v_p{}^2 l_p{}^2 = l_p{}^3/l_m{}^3 = 50^3$, and $F_{d,\ p} = 0.27 \times 50^3$ N.

We may also calculate the power, P, which is equal to the product of force and velocity ($ML^2T^{-3} = MLT^{-2}.\ LT^{-1}$), $(P_p/P_m) = (F_{d,\ p}/F_{d,\ m})\,(v_p/v_m) = l_p{}^{7/2}/l_m{}^{7/2}$.

If the water depth, assumed to be finite and of a depth d, were also added to the initial hypothesis for a dimensional analysis, the only modification would be one additional dimensionless Π-group, l/d, which would modify equation (13.14) to give:

$$F_d/v^2 l^2 = f\left(\frac{\mu}{\rho v l},\ gl/v^2,\ \frac{l}{d}\right). \tag{13.18}$$

For large values of l/d, it might be possible that it is the depth d which might be more important in the dimensionless gravity term. The following is a possible re-arrangement:

$$F_d/\rho v^2 l^2 = f\left(\frac{\mu}{\rho v l},\ gd/v^2,\ \frac{l}{d}\right). \tag{13.19}$$

There is a full description of using models of boats, and the problem of simultaneously matching Re and Fr in [2].

13.3 Dimensional analysis of a water sport

Excluding hydroplaning boats, how fast can a boat move? Longer boats generally move faster than shorter boats, so it is likely that the length of the boat, l, determines the top speed. The density of water might also influence the speed. However, using only v (the velocity, LT^{-1}), ρ (density, ML^{-3}) and l (L), we cannot form any dimensionless groups. So, we need to add, at least, one more variable. Viscosity (μ) is irrelevant, because the Reynolds number for the motion of boats is large (see table 1.2 for the dimensions of viscosity); even for a small boat of length 5 m, moving at 2 m s^{-1} (4 knots), Re ($\equiv \frac{\rho v l}{\mu}$) $\sim$ 500 cm. 200 cm $s^{-1}/10^{-2}$ cm^2 s^{-1} $\sim 10^7$. With such a high Reynolds number, the flow is turbulent, and therefore independent of viscosity. Surface tension is also irrelevant, because boats are much longer than a surface ripple wavelength (roughly a few cm). The missing variable, needed to form nondimensional groups to model the motion of a surface boat, is the acceleration due to gravity (LT^{-2}). From v, ρ, g and l, we can form one dimensionless group (there are four variables composed of three base dimensions), which is v^2/gl, also called the Froude number.

The critical Froude number, which determines the maximum boat speed (and the speed of bipeds—see section 14.2), is a dimensionless group of variables. As usual, we assume that the constant of proportionality is, of order, unity: $v \sim \sqrt{gl}$. We use this expression, because when a boat moves through water there are waves generated by that motion. The moving boat generates a wake, and it rides on one of those self-generated waves. This bow wave is a gravity wave with phase velocity similar to the boat velocity: $v_{\text{phase}} \sim v_{\text{boat}}$. Because $v^2_{\text{phase}} = \omega^2/k^2$, the dispersion relation tells us that:

$$v^2\text{boat} \sim \omega^2/k^2 = g/k = g\,\lambda^*, \tag{13.20}$$

where $\lambda^* \equiv 1/k = \lambda/2\pi$. So, the wavelength of the waves is $\sim v^2_{\text{boat}}/g$. The other length in this problem is the boat length, l. So, we may interpret the Froude number as

$$\text{Fr} \equiv v^2\text{boat}/g\,/l \sim \frac{\text{wavelength of bowwave}}{\text{length of boat}}. \tag{13.21}$$

Interesting physics occurs when a dimensionless number is near unity—the regimes of approximation are merging, and the approximations are no longer appropriate. In this case, when Fr ~ 1, the wave height changes significantly in a distance λ^*; if the boat's length l is comparable to λ^*, then the boat rides on its own wave and tilts upward. It then presents a larger cross-section to the water, and the drag force becomes larger. So, the top speed is given by

$$v_{\text{boat}} \sim \sqrt{gl}. \tag{13.22}$$

For a small motorboat, with length $l \sim 5$ m, this speed is roughly 7 ms^{-1}, or 15 mph, or (14 knots). Boats can go faster than the nominal top speed given in equation (13.22), but it takes a great deal of power to fight the drag force, which is why police speedboats have huge engines. This is the same problem as for cars on a motorway (see section 8.2.4). The faster you go in your non-aerodynamic family car, the greater the drag force, and the more petrol you consume. However, if you drive a well-designed Lamborghini, you travel faster with greater efficiency.

References

[1] Santiago J G 2019 *A First Course in Dimensional Analysis* (Cambridge MA: MIT Press)
[2] Fox R W, McDonald A T and Pritchard P J 2015 *Introduction to Fluid Mechanics.* 9th edn (Hoboken, NJ: Wiley)

Chapter 14

The great principle of similitude in biology and sport

'The harmony of the world is made manifest in Form and Number, and the heart and soul and all the poetry of Natural Philosophy are embodied in the concept of mathematical beauty.'

D'Arcy Wentworth Thompson (1860-1948)

Natural history has demonstrated that biological diversity is largely a matter of body size, which is seen to vary over about ten orders of magnitude in the natural world—from bacteria to the great whales. This vast range in size influences all biological structures and the rates of physiological processes from cellular metabolism to population dynamics. Observation has shown us that the dependence of a biological variable, Y, on body mass, M, is typically characterized by a scaling law of the form:

$$Y = Y_0 M^b, \tag{14.1}$$

where b is the scaling exponent, and Y_0 a constant that is characteristic of the kind of organism under study. This simple rule is an example of an allometric law; that is, a rule derived from the study of the relationship of body size to: shape, anatomy, physiology and behaviour. Such rules were first outlined by Otto Snell in 1892, and by D'Arcy Thompson in *On Growth and Form* [1]. If, as originally thought by Galileo, such relations reflect geometric constraints, then b should be a simple multiple of one-third (the property of interest will be proportional to the volume (V) of the organism; that is $Y \propto V$, and so Y should scale as $l^{1/3}$, where l is a single geometric length dimension, L). However, most biological phenomena scale as quarter powers, rather than third powers of body mass; for example, metabolic rates, B, of organisms scale as $M^{3/4}$; rates of cellular metabolism, heartbeat and

doi:10.1088/978-0-7503-3655-0ch14

maximal population growth scale as $M^{-1/4}$; and times of blood circulation, embryonic growth and development, and life-span scale as $M^{1/4}$. Sizes of biological structures scale similarly; for example, the cross-sectional areas of mammalian aortas and of tree trunks both scale as $M^{3/4}$. No general theory explains the origin of these laws. Presently, they are merely observations.

The best known of these allometric relations is Kleiber's law, named after the Swiss biologist Max Kleiber (1893–1976) for his work on the metabolism of different animals in the early-1930s. The law is basically the observation that for the vast majority of animals, their metabolic rate scales to the ¾-power of the animal's mass. Thus, if q_0 is the animal's metabolic rate, and m the animal's mass, then Kleiber's law states that $q_0 \sim m^{3/4}$, see equation (14.1). Kleiber plotted the metabolic rate of a huge variety of animals against the mass of those animals, and he obtained a broadly linear relationship. The range of animals is quite extraordinary; from the smallest of mice to the great whales, with metabolic rates varying over six orders of magnitude. A measure, B, of metabolic rate is simply the heat lost by a body in a steady inactive state, which can be expected to be dominated by the surface effects of sweating and radiation. Symbolically, therefore, one expects $B \propto W^{2/3}$. Kleiber's analysis showed that metabolic rate does indeed scale; that is, all animals lie on a single line in spite of the fact that an elephant is not an oversized mouse, or a whale an oversized minnow. However, the slope of the best-fit curve is closer to 3/4 than to 2/3, indicating that effects other than the pure geometry of surface dependence are at work. The exact value of the exponent in Kleiber's law is unclear, in part because there is currently no completely satisfactory theoretical explanation for the law. It is an observation, in need of a theory [2, 3].

In his book *Dialogo sopra i due massimi sistemi del mondo* (Dialogue Concerning the Two Chief World Systems) of 1632, Galileo Galilei made a simple and clear comment on the effects of scaling up the dimensions of a physical object; that is, on the nature of the principle of similitude. Galileo realized that if one simply scaled up its size, the weight of an animal would increase significantly faster than its strength, causing it ultimately to collapse upon itself—it would fail as an engineered structure. As Galileo says, '... *you can plainly see the impossibility of increasing the size of structures to vast dimensions ... if his height be increased inordinately, he will fall and be crushed under his own weight.*' The simple scaling up of an insect to an enormous size is thus a physical impossibility—much to the chagrin of writers of science fiction. Clearly, to create a giant of something, one '*must either find a harder and stronger material ... or admit a diminution of strength*'; a fact long known to architects. It is remarkable that so many years before its deeper significance could be appreciated, Galileo had investigated one of the most fundamental questions of Nature: namely, what happens to a physical system when one changes scale? Today, this is the central question for quantum field theory, phase transition theory, the dynamics of complex systems and attempts to unify the four forces of Nature.

For similar structures; that is, structures having the same physical characteristics such as shape, density, or chemical composition, Galileo perceived that weight, W, increases linearly with volume, V, whereas strength increases as a cross-sectional area, A. Since for similar structures $V \propto l^3$ and $A \propto l^2$, where l is some characteristic length (such as the height of a structure or person, or the length of an airplane fuselage, a submarine, or of an insect's body), we conclude that:

$$\frac{\text{Strength}}{\text{Weight}} \propto \frac{A}{V} \propto \frac{1}{l} \propto \mathrm{W}^{-1/3}. \tag{14.2}$$

Thus, as Galileo noted, smaller animals '*appear*' stronger than larger ones. Arguments of this sort were used extensively during the late-19th Century to understand the grosser features of the biological world; indeed, the general size and shape of animals and plants can be viewed as Nature's way of responding to the constraints of gravity, surface phenomena, viscous flow and the like. For example, one can understand why man cannot fly under his own muscular power, and why small animals leap as high as larger ones, and so on. Typically, deviations from a simple geometrical, or kinematical similitude, reflect the dynamics of the system; that is, one can view deviations from naive scaling as a probe of the dynamics.

14.1 Scaling of flight

A particularly interesting scaling argument arises if we consider the flight of birds. From dimensional analysis (see section 5.2.2) we argued that the drag force, F_{d} and the lift force, F_{L} are given by,

$$F_{\mathrm{d}} = F(l, b, v, \mu, \rho, \theta) \text{ and } F_{\mathrm{L}} = F'(l, b, v, \mu, \rho, \theta), \tag{14.3}$$

where l is the wing length, b is width of the wing, v is the speed of flight, μ is the viscosity of air, ρ is the density of air and θ is the angle of attack (dimensionless). The dimensions of these variables may be found in table 5.2. Using the Buckingham Π-theorem to analyse this problem, we have seven physical variables involving three base dimensions, which tells us that we need four nondimensional Π-groups to define the problem. In addition to the Reynolds number (Re) and the drag/lift coefficient, we may choose l/b and θ as Π-groups. Thus, by Buckingham's Π-Theorem[1],

$$F_{\mathrm{L}}/\rho l b v^2 = G\left(\mathrm{Re}, \frac{l}{b}, \theta\right). \tag{14.4}$$

The physical dimensions of birds are found to be geometrically similar [4]. It is seen that $l \propto m^{1/3}$, and $b \propto m^{1/3}$ so the projected area of the wings is proportional to $m^{2/3}$, and the wing loading (loaded weight per wing area) increases as $m/m^{2/3} = m^{1/3}$. Indeed, if the wing loading for a huge range of birds, insects and aircraft is plotted as a function of the body mass, a straight line with a slope of 1/3 is seen to be a remarkably good fit (figure 2 in [4]).

[1] In chapter 5, we analysed the lift force with a six-variable model (equation 5.9); this led to three nondimensional Π-groups. Here, to give a closer approximation to reality, we use a model of seven variables (equation (14.3)); that is, we include a wing length and a wing breadth. This means there are now four nondimensional Π-groups.

As birds display a strong similitude, the ratio of wing-width to wing-length, b/l, is a constant. The wing angle of attack is also near constant, so we can reduce equation (14.4) to

$$F_L/\rho l^2 v^2 = G_1(\mathrm{Re}), \tag{14.5}$$

where we have assumed b to be proportional to l. Over the range of Reynolds numbers for bird flight (10^3 to 10^5), it is found that the lift coefficient varies only slightly with the Reynolds number, so we may write

$$F_L/\rho l^2 v^2 = k_1, \tag{14.6}$$

where k_1 is a constant. For a bird to fly, the lift force must, at least, balance the weight of the bird ($F_L = mg$). Thus, $mg/\rho l^2 v^2 = k_1$, and we may make the prediction that $v \sim k_2\sqrt{m/l^2}$.

The ratio m/l^2 is called the span loading, and the above scaling argument tells us how it is related to the cruising speed of a bird; there is good agreement with the calculated values for many different birds [4]. From self-similarity, $m \propto l^3$, and we generate the relationship, $v \sim l^{1/2} \sim m^{1/6}$; implying that larger birds fly faster than smaller birds.

Birds and airplanes can change the angle of attack of their wings as needed. They fly nose-up, with a high angle of attack, when they have to fly slowly, or have to make a sharp turn; they fly nose-down when speeding or diving. But everything that flies in the air uses a similar angle of attack in long-distance cruising—about 6°. At higher angles of attack the aerodynamic drag on wings increases rapidly; at smaller angles, wings are underutilized. Since wings have to support the weight of an airplane, or a bird against the force of gravity, the lift, F_L, must, at least, equal the weight, w. The lift is proportional to the wing area, S ($\sim lb$) and to ρv^2; and the weight is

$$w = 0.3\,\rho v^2 lb. \tag{14.7}$$

(The 0.3 is related to the angle of attack in long-distance flight, for which the average value of 6° has been adopted [4].)

Equation (14.7) is typical of the class of relations defined by equation (14.1), and can be used in several ways ([4] contains a very readable, and informative description of natural and man-made flight). For example, a Boeing 747 has a wing area of 511 m^2, and cruises at a speed of 250 m s^{-1}, at an altitude of 12 km (40 000 feet), where the air density (ρ) is only one-fourth its sea-level value of 1.25 kilogram per cubic metre. Using $\rho = 0.3125$ kilogram per cubic metre, $v = 250$ m s^{-1}, and $S = 511$ m^2, we calculate from equation (14.7) that $w = 2\,990\,000$ N; that is, 300 000 kg (1 N $\approx$ 0.1 kg). That is indeed the weight of a 747 at the midpoint of an intercontinental flight. At take-off, it is considerably heavier, due to the fuel. Equation (14.7) may also be used in designing a hang-glider. Taken together, the pilot and the wing weight are about 1000 N (100 kg). So, if you wish to fly as fast as a sparrow (20 miles per hour, 9 m s^{-1}), you need wings with a surface area of 33 m^2. Given that most birds fly at relatively low altitudes, the density of air can be replaced by the sea-level value, and then if we divide both sides by the wing area S, we have $w/S = 0.38v^2$. This scaling relationship tells us that the greater a bird's wing loading, w/S, the faster the bird must fly. Within the approximations we are using here, sea-level cruising speed depends on wing loading only—no other quantity is involved.

The scale relations described above are applicable whenever weights and supporting surfaces, or cross-sectional areas are involved. As mentioned, Galileo Galilei wrote the first treatise on this subject, asking himself why elephants have such thick legs. The answer is that the larger an animal gets, the more crucial the strength of the legs becomes; the stress on leg bones increases as the cube root of weight. This is the same problem that engineers face when they design bridges or skyscrapers, which would give way under their own weight were they not reinforced by steel girders. Another good example is that of walking barefoot on a stony beach. Walking on gravel is an uncomfortable experience for adults, but less so for small children. An adult who is twice as tall and eight times as heavy compared to a child must support themselves on feet whose surface area is only four times larger than that of the child's feet. Thus, the adult's foot loading is twice that of the child. These scale relations are, however, not hard-and-fast rules. Most birds are not exact scale models of others—evolution brings about such deviations from similarities. We must also allow some latitude for deviations to fit designers' visions and nature's idiosyncrasies (ecological niches). The range of variability seen in Nature is huge, but the margins permitted by the laws of scaling are finite [4].

14.2 Walking and running with dinosaurs

When compared to electromagnetism, gravity is not a strong force. It poses no immediate danger to life; unless you fall from a tall building. Yet all life on Earth has been exposed to the gravitational force-field of the Earth, and so every living thing is the result of evolution in this gravity-field. The effects of gravity on our bodies are slow-acting; for example, our breathing involves the use of muscles to vary the size of the thoracic cavity of our bodies. This mechanism has evolved in the Earth's gravity-field, and our intercostal muscles have evolved to be able to deal with the air pressure and gravitational pull of our planet. If the air pressure increased significantly, our descendants would need to evolve by developing more powerful muscles to make it easier to breathe, and probably to enlarge the size of the lungs in response to this change.

Similarly, gravity is important when it comes to the manner in which we have evolved to move about unaided. Moving, or accelerating from walking to walking more quickly, or even running is something we take for granted. You may readily double your speed, but your pattern of bodily movements remains the pattern we call walking. Then, suddenly, you break into a run; now there is a different pattern of movement. How is this possible? Why do we use one technique of movement at low speeds, and a dramatically different technique at high speeds? The change is made at a very predictable speed; about 2.5 m s^{-1} for average-sized adults. Walking involves each foot being on the ground for slightly more than half the time, so that there are several stages of walking; for example, a stage when both feet are on the ground simultaneously. While a foot is on the ground, its associated leg remains fairly straight; therefore, the trunk of the body is highest when the leg is straight and vertical. The trunk is lowest when both feet are on the ground. In running, each foot is on the ground for less than half the time, so that there are also stages—when one or both feet are on, or off the ground. When running, the person is travelling in a series of leaps and must therefore be highest in mid-leap. The runner's supporting leg is bent as the trunk passes over the supporting foot. The trunk is lowest at this stage.

We can make an attempt at explaining why people walk only at fairly low speeds, by formulating a simple model. The model runner sets one foot down at the instant he lifts the other. While each foot is on the ground, he/she keeps the leg rigidly straight, so that his/her hip joint moves along an arc of a circle centred on the foot (see figure 14.1). The legs are sufficiently light, compared to the trunk, for their mass to be ignored. Thus, the runner's centre of mass occupies a fixed position in their trunk, and moves along arcs of the same radius as the path of the hip joint. This radius is the length, l, of the leg. Let a man's walking speed be v; Alexander [5] presented a simple model of running and walking—the two basic gaits of bipeds. A walking biped typically fixes the knee of the leg that touches the ground, and so the motion is like a rotation about the foot contact point. Think of this as an inverted pendulum—a fixed-rod pendulum, with a pivot on the ground and the oscillating bob centred at the runner's pelvis (figure 14.1). The straight leg, of length l, being the wire of the conventional pendulum.

When walking, see figure 14.1, a vertical force balances the motion of the inverted pendulum, and we may write (using Newton's second law),

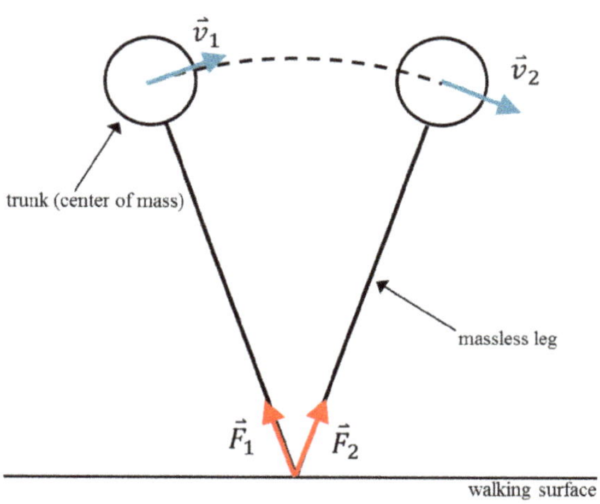

Figure 14.1. Centre of mass on a 'massless' model leg travelling along the trajectory followed by the pelvis in walking—it follows the motion of an inverted pendulum. Velocity vectors are shown perpendicular to the ground reaction force ($\mathbf{F}_1$ and $\mathbf{F}_2$) at time 1 and time 2. The centripetal acceleration of the walker's pelvis is mv^2/l. The inverted pendulum motion allows the reaction force (**F**) to be converted into forward momentum. This 'inverted pendulum' image has been obtained by the author from the Wikimedia website where it was made available under a CC BY-SA 3.0 license. It is included within this article on that basis. It is attributed to Jenna Fair.

$$N - mg = ma = -mv^2/l, \tag{14.8}$$

where N is the normal force exerted by the ground and $-mv^2/l$ is the centripetal acceleration. Since N is strictly positive (upwards), the maximum speed for walking is then limited by $N = 0$, where $mv^2/l = mg$. That is, a point moving with speed, v, along an arc of a circle of radius, l, has an acceleration, v^2/l, toward the centre of the circle. Thus, the centre of mass of the walker has an acceleration v^2/l toward the supporting foot. Since the athlete cannot pull himself down, but can only fall under gravity, their downward acceleration, v^2/l, cannot exceed the acceleration of free fall, g, that is, $v^2/l \leqslant g$, which by transformation gives: $v \leqslant \sqrt{gl}$. Thus, the model athlete cannot walk faster than $\sqrt{gl}$.

On Earth, g is about 10 ms^{-2}, and leg length (from the hip joint to the ground) is typically 0.9 m. In an adult, therefore, the maximum walking speed v is about 3 m s^{-1}. This is only a little faster than the approximately 2.5 m s^{-1} at which real people break into a run. Children have shorter legs than adults, and the theory correctly predicts that they will have to begin running at lower speeds. This model also seems to explain why accident victims sometimes find that they can move faster than they expected on crutches: someone on crutches moves in arcs about 1.4 m instead of the usual 0.9 m for walking. If the running speed is at the critical value, defined by $v^2/l \leqslant g$, then the force on the leg is zero because the gravity force balances the centripetal force. Just above this speed, the incipient runner's foot leaves the ground and, at some point neither foot is in contact with the ground—the athlete floats. This is the essential difference between walking and running. Moving even faster requires longer times and distances where the body is not in contact with the ground.

The criteria for the inequality, $v^2/l \leqslant g$, can also be expressed as Fr $\leqslant 1$, where Fr is the nondimensional Froude number (see section 4.1); defined by Fr $\equiv v^2/gl$. The Froude number, named for the English naval architect, William Froude, is a ratio of the kinetic and potential energies of motion. Thus,

$$1/2Fr = v^2/2gl = 1/2mv^2/gml = \frac{\text{Kinetic energy}}{\text{Potential energy}}. \tag{14.9}$$

Although one might imagine that as bipeds evolved from quadrupeds, there should be a certain similarity in the biodynamics of motion, quadrupeds are strikingly dissimilar to bipeds, with very different gaits. As Alexander describes [5]; quadrupeds have three basic gaits: walk, trot and gallop. However, we will here hypothesise that the idea of a critical Froude number dividing walking from running in bipeds may also be applicable as a criterion driving the different gaits of quadrupeds. If this is born out, then the Froude number could be a general characterization of all animal gait. And by the principle of similitude, this criterion would apply to extinct animals as well as those living today.

Consider a dimensional analysis of the maximum running speeds of animals. The variables to be considered and their dimensions are given in table 14.1.

This dimensional analysis, first demonstrated by Alexander in 1983 [5], is not based on geometric similitude—quadrupeds and bipeds are dissimilar. Instead, the idea is to seek to know if a nondimensional number is a strong indicator of the

Table 14.1. Variables for biodynamics.

Variable	Dimension	Description
v	LT^{-1}	Velocity of a specific gait of movement (e.g. running)
L_{stride}	L	Distance travelled in a single complete cycle of leg movements
L_{leg}	L	Distance from hip to ground
g	LT^{-2}	Acceleration due to gravity

biomechanics of all animals. This in itself will demonstrate that it is possible to use the principle of similitude to examine the fossil remains of dinosaurs, and determine how fast they ran, or walked, or flew. The principle of uniformitarianism; that is, the present being the key to the past is valid, provided the principle of similitude is upheld. Does the inverted rod-like pendulum have any bearing on the running gait of a *Tyrannosaurus rex* (see figure 14.2)?

These two length directions in table 14.1 are orthogonal; in that one is the length of a stride, and the other is the length of the limb making that stride. The leg length will be a fixed parameter for each animal, while length of stride will be determined by running speed. Alexander [5] commented that hip height is typically nearly the same as height to the centre-of-mass of a running animal; a condition imposed by evolution on a planet with an acceleration due to gravity of about 10 m s^{-2}. We hypothesise that the velocity of an animal is given by

$$v = f\left(L_{\text{stride}}, L_{\text{leg}}, g\right). \tag{14.10}$$

As with the analysis of the pendulum (see section 6.3), we exclude the mass of the animal, because mass does not determine the period of a pendulum. The dimensional analysis yields

$$\left(v/\sqrt{gL_{\text{leg}}}\right) = f\left(L_{\text{stride}}/L_{\text{leg}}\right). \tag{14.11}$$

We have two nondimensional terms, one of which is the square-root of a Froude number (in terms of the length of the leg). This gives

$$L_{\text{stride}}/L_{\text{leg}} = F\left(v/\sqrt{gL_{\text{leg}}}\right) = F\left(\sqrt{\text{Fr}}\right). \tag{14.12}$$

Graphs from Alexander (1996) [5], which may be found in Santiago [6], show a comparison between data for stride length (0.5 to 4.5 m) versus velocity (0.5 to 8 m s^{-1}) for three animals: man, dog and camel. The three data sets are fairly linear, albeit with different slopes and y-intercepts. The two lines for dogs and camels are almost parallel (slope ≈ 1/4), but the line through the human data has a much higher slope (slope ≈ 1/2). When all the data sets are plotted as $L_{\text{stride}}/L_{\text{leg}}$ versus $\sqrt{\text{Fr}}$, one obtains a single straight line (even when data for rhinos and kangaroos is added). This straight line has a slope ≈ 4/3, and a y-intercept of 0.82, which characterizes the function, F, in equation (14.12).

Figure 14.2. *Tyrannosaurus* and *Triceratops* at Los Angeles Natural History Museum. The *T. rex* can clearly be imagined to run on its two powerful hind legs. This '*Tyrannosaurus* and *Triceratops* at Los Angeles, Natural History Museum' image has been obtained by the author from the Wikimedia website where it was made available under a CC BY 2.0 license. It is included within this article on that basis. It is attributed to Allie Caulfield.

From television shows to Hollywood blockbusters, dinosaurs are perennial favourites for inspiring people's curiosity. And topping the fascination list is *Tyrannosaurus rex*, the six-ton giant that reigned over the Upper Cretaceous Period, 65 million years ago, moving around on hind legs more than 2.5 metres long, devouring less fortunate (and slower moving) creatures. Although the fossil record provides clues about some aspects of dinosaur existence, insights into many others must be teased out of fossils and petrified footprints, but any questions lack direct answers. The successful collapse of data of the biodynamics from a wide-range of extant species onto a single model fit tends to suggest that extinct species might also be accommodated on such a model fit. There is no reason to believe that the force of gravity of planet Earth has changed much over the last few billion years. Consequently, to estimate the running and walking speed of dinosaurs we need only have the dimensions of their legs, and the model curve mentioned above for men, camels, dogs, etc. The dimensions of dinosaur limbs are readily available from the fossil record. Analysis of such fossils (and the tracks of petrified footprints, which give L_{stride}) provide the data in table 14.2 (data taken from Santiago [6]). The analysis is performed using

$$L_{\text{stride}}/L_{\text{leg}} = F\left\{v/\sqrt{g}L_{\text{leg}}\right\} = m.\left(v/\sqrt{g}L_{\text{leg}}\right) + b, \tag{14.13}$$

with $b = 0.82$ and $m = 4/3$. (A fascinating discussion of the biodynamics of dinosaurs is to be found in [7].)

Disappointingly, the running speed of large dinosaurs was fairly low. *Tyrannosaurus rex*, for example, would have had difficulty catching a human casually jogging at 5 mph. Even small carnivores such as velociraptors are estimated to have

Table 14.2. Size of some dinosaurs (taken from [6]).

Dinosaur	Leg length (m)	Estimated speed (m s^{-1})
Large Therapod (*Tyrannosaurus rex*)	2.0	2.2
Large Sauropod	3.0	1.0
Ornithopods	0.14–1.6	4.3–4.8

run at only about 8 mph (3.5 m s^{-1}). The fastest of these extinct giants were the ornithopods; some of whom managed to get to 11 mph. One presumes that there was no evolutionary pressure back in the Mesozoic that required the carnivores to improve their running speed—there was plenty of food that was slower than they were. Most healthy adults can manage running speeds of 15–20 mph, and the human world speed record is roughly 25 mph (11.2 m s^{-1}); thereby leaving a *T. rex* in their dust.

14.3 Constructing the best First-VIII

Rowing shells, that is, eights, fours, and double and single skulls are used in inter-varsity competitions, at the Olympic Games, and rowing competitions (see figure 14.3 for such eights). Such competitions have boats containing one, two, four or eight rowers. One immediately poses two questions:

- Does a one-person boat go faster than an eight-person boat?
- Does the weight of the rower influence the speed of the boat? That is, would it be better to have a lightweight or heavyweight rower?

Determining the maximum speed of a rowing eight is a classic example of the simplicity and elegance of dimensional analysis, and was originally analysed by McMahon (1971) [8].

This is quite a complex problem, and to make any progress one needs to make three significant assumptions.

- Boats with different numbers of rowers are geometrically similar to each other (they are all long and thin).
- The submerged volume of a boat, per rower is the same, and given by V_0.
- The power supplied by a rower is always the same, and equal to P_0.

Figure 14.3. The men's Second-VIII of Trinity Hall, Cambridge, 'bumping' the men's Second-VIII of Clare College in the Lent Bumps 2011. This 'Trinity Hall II chasing Clare II' image has been obtained by the author from the Wikimedia website where it was made available under a CC BY-SA 3.0 license. It is included within this article on that basis. It is attributed to William M. Connolley.

Before proceeding to analyse the nature of rowing in an eight, as compared to a single-person scull, we need to understand which physical processes govern the maximum speed of the boats. The dominant drag force is that of friction with the water. Note that if the boat were larger, or the speed greater, the typical Froude number would also be large. In such a case, the drag would have been governed by the generation of surface waves. For a typical (rowing) boat, the Froude number is small, and hence the effect of surface gravity waves is negligible. Since the motion of the water is turbulent (there are vortices behind the boat), the frictional drag force is given by

$$F_d \sim \rho v^2 l^2. \tag{14.14}$$

This can be seen from the dimensional analyses given earlier in this volume; the only force we can construct from the density of water ρ, the speed of the boat v and its geometric length dimension l is, $\rho v^2 l^2$. The power lost to drag is therefore,

$$P_{\mathrm{d}} = F_{\mathrm{d}} v \sim \rho v^3 l^2, \tag{14.15}$$

just as it would be for a car travelling on a highway (see section 8.2.4). Since we assume geometric similarity, the size of the boat l will be proportional to the submerged depth, which itself should be proportional to the cubic root of the submerged volume. Thus, $l \sim V^{1/3} \approx (NV_0)^{1/3}$, where N is the number of rowers. The total power supplied by the rowers, P_{rowers}, should be proportional to the number of rowers, N, each supplying a power of P_0. At steady-state, the power of the rowers should be the same as the power lost to drag. We may write, $P_{\mathrm{rowers}} \approx P_{\mathrm{d}} \rightarrow NP_0 \sim \rho v^3 N^{2/3} V_0^{2/3}$ or $v \sim N^{1/9} P_0^{1/3} / \rho^{1/3} V_0^{2/9}$. In other words, the speed of a boat with N rowers scales as $N^{1/9}$. The time it would take them to finish a race, t, will therefore scale as: $t \propto N^{-1/9}$. Different type boats will, of course, have different constants of proportionality since V_0 and C_{d} (the drag coefficient, which is the constant of proportionality in the drag equation—equation 8.4) are somewhat different. But for a given type of boat, we should see this scaling behaviour. What happens if there is a coxswain? In such a case, there is an extra person who does nothing (from a physical point of view), except add weight and volume. This implies the power supplied by the N rowers is the same, but the drag they need to overcome is larger:

$$NP_0 \sim \rho v^3 (N+1)^{2/3} V_0^{2/3}, \quad \text{or } v \sim N^{1/3} (N+1)^{-2/9} P_0^{1/3} / \rho^{1/3} V_0^{2/9}.$$

Let us now consider a more formal solution to this problem of the biomechanics of rowing. Assume a boat has N rowers, and each rower has the same mass, M_0, and can put the same power, P_0 into propelling the boat; the density of the water is ρ, and the velocity of the boat is v. We hypothesise that the velocity of the boat is given by $v = f(NM_0, NP_0, \rho)$. These variables are detailed in table 14.3.

There is no need to use the Buckingham Π-theorem; the solution will have the form $v = c\, D$, where c is a constant and D is a quantity defined by $D = (NM_0)^x (NP_0)^y (\rho)^z$, such that $[D] = [v]$. The problem is solved via the Rayleigh algorithm.

Thus, $[D] = [v] = [NM_0]^x [NP_0]^y [\rho]^z$, and the dimensional equation is

Table 14.3. Variables for rowing.

Variable	Dimension	Description
N	–	Number of rowers
NM_0	$[N\ M_0] = \mathsf{M}$	Total combined mass of rowers
NP_0	$[N\ P_0] = \mathsf{ML^2\ T^{-3}}$	Total combined power output of rowers
ρ	$\mathsf{ML^{-3}}$	Density of water
v	$\mathsf{LT^{-1}}$	Velocity of boat

$$\mathsf{LT^{-1}} = (\mathsf{M})^x(\mathsf{ML^2T^{-3}})^y(\mathsf{ML^{-3}})^z = \mathsf{M}^{x+y+z}\mathsf{L}^{2y-3z}\mathsf{T}^{-3}. \tag{14.16}$$

The solutions are: $x = -2/9$, $y = 1/3$ and $z = -1/9$.

Thus,

$$D = (NM_0)^{-2/9}(NP_0)^{1/3}\rho^{-1/9}, \text{ and } \Pi = \frac{v}{D} = v/(NM_0)^{-2/9}(NP_0)^{1/3}\rho^{-1/9}, \tag{14.17}$$

which is dimensionless. As stated earlier, $\Pi = c$ where c is a constant, and thus,

$$\begin{aligned} v &= c(NP_0)^{1/3}/(NM_0)^{-2/9}\rho^{-1/9} = c\left\{(NP_0)^3/(NM_0)^2\rho\right\}^{1/9} \\ &\approx (N^3/N^2)^{1/9}. \end{aligned} \tag{14.18}$$

From this result, we may now attempt to answer our original two questions. Firstly, we note from the result that $v \propto N^{1/9}$, from which we infer that an eight-person boat (an eight) will go faster than a one-person skull. Secondly, we suppose that $P_0 \propto M_0$ or $P_0 = kM_0$. Then,

$$\begin{aligned} v &= c\{(NP_0)^3/(NM_0)^2\rho\}^{1/9} = c\{N(kM_0)^3/M_0{}^2\rho\}^{1/9} \\ &= c(Nk^3M_0/\rho)^{1/9} \\ &= (ck^{1/3}/\rho^{1/9})(NM_0)^{1/9}, \end{aligned} \tag{14.19}$$

and since $ck^{1/3}/\rho^{1/9}$ = a constant, we note that $v \propto (NM_0)^{1/9}$, and conclude that size is of advantage to a rower. See figure 14.4 for a comment on the verification of this result.

14.4 How much can you lift?

Having used dimensional analysis to demonstrate that a *Tyrannosaurus rex* would not be able to catch a moderately fit jogger, and how one constructs the best rowing eight, let us now look at how much a weightlifter might be able to lift.

The insight needed to relate the characteristics of a weightlifter to the weight the weightlifter can lift, is that the lifted weight will scale with the muscle stress (σ_m) and the cross-sectional area of those muscles. The significant assumption being made is that all weightlifters are approximately alike; that is, we can assume geometric similitude. The other assumption is that the muscles in all weightlifters (there are

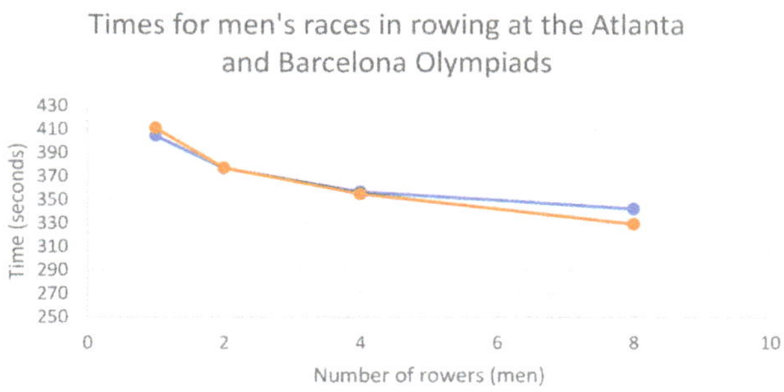

Figure 14.4 The times of the men's finalists in the single skulls class, coxed doubles, coxed fours and coxed eight class at the Atlanta (blue) and Barcelona (orange) Olympiads. The slope of the data for the $N = 2$, 4 and 8 rowers is, of order, 6 s per rower. This data is, however, only of limited use in verifying the $N^{1/9}$ scaling. N varies by a factor 8, and the 9th root of 8 is 0.9; that is, there should be a 10% decrease in time for these different sized boats. We can see more than a 10% fall in race times. So, we need more experiments, performed under identical (temperature, wind speed, etc) conditions.

various categories of lifters—from lightweight to heavyweight) have the same physical properties—particularly, body density (ρ).

Given that we are assuming geometric similitude among the weightlifters, we may use the Rules of Thumb given in chapter 7 to say that the individual body characteristics: volume, weight and cross-sectional area of muscle may be scaled with the same geometric parameter. Here, we will select the weightlifters' heights, h, to be this single geometric scaling parameter (the weight has to be lifted, against gravity (g), above the lifter's head); having defined this parameter, we can now say that the weight of the weightlifter, which defines the class in which he/she lifts will scale as h^3.

We are interested in the weight the weightlifters can lift, m_l, and we hypothesise for this dependent variable: $m_l \propto (h, \sigma_m, g, \rho)$. The variables involved in this proposed model are described in table 14.4. We see that there are five variables, involving three base dimensions; and so, using the Buckingham Π-theorem, there are two nondimensional groups to be identified.

We will construct the first dimensionless group from the dependent variable, and the second from the muscle stress, with h, g and ρ being the repeat variables. Thus, we may write for the dimensional equation:

$$\Pi_1 = m_l[h]^x[g]^y[\rho]^z = \mathsf{M}(\mathsf{L})^x(\mathsf{LT}^{-2})^y(\mathsf{ML}^{-3})^z = \mathsf{M}^{z+1}\mathsf{L}^{x+y-3z}\mathsf{T}^{-2y}. \tag{14.20}$$

The solutions are: $x = -3$, $y = 0$ and $z = -1$. Thus, $\Pi_1 = m_l/h^3\rho$; which is essentially a ratio of the mass of the weight lifted and the mass of the weightlifter. The second Π-group is found in an analogous manner.

$$\Pi_2 = \sigma_m[h]^x[g]^y[\rho]^z = \mathsf{ML}^{-1}\mathsf{T}^{-2}(\mathsf{L})^x(\mathsf{LT}^{-2})^y(\mathsf{ML}^{-3})^z = \mathsf{M}^{z+1}\mathsf{L}^{x+y-3z-1}\mathsf{T}^{-2y-2}. \tag{14.21}$$

Table 14.4. Variables for the biomechanics of weightlifters.

Variable	Dimension	Description
m_l	M	Weight lifted
h	L	Height of lifter
σ_m	$ML^{-1}T^{-2}$	Maximum muscle stress
g	LT^{-2}	Acceleration due to gravity
ρ	ML^{-3}	Body density of lifter

This time, the solutions are: $x = y = z = -1$, and $\Pi_2 = \sigma_m/h\rho g$. The Buckingham Π-theorem allows us to write $\Pi_1 = f(\Pi_2)$, and so the final result is:

$$m_l/h^3\rho = f(\sigma_m/h\rho g), \text{ and } m_l = h^2\sigma_m/g. \tag{14.22}$$

This is a force divided by the acceleration due to gravity, which may be rearranged to give a statement of Newton's second law: $F = m_l g = h^2\sigma_m$ = (mass) × (acceleration). The lifters are seeking to lift a weight against the force of the Earth's gravity. In the dimensional analysis, we defined the lifters by their body heights (a dimension of L). It is a rule of dimensional analysis (see chapter 7) that one should avoid redundant variables, because this leads to superfluous nondimensional groups of variables (Π-groups) and leads to less physical insight. Given a single length dimension for the lifters, and assuming geometric similitude, the body weight of the lifters will scale as h^3.

Although not included specifically in the variables for the dimensional analysis, the mass of the weightlifters (m_0) is given by $m_0 = c_0\rho h^3$. Here, c_0 is a nondimensional constant that arises by assuming geometric similitude—and so was not included in the initial hypothesis. Our analysis yielded the following two dimensionless groups: $m_l/h^3\rho = F_1(\sigma_m/h\rho g)$, where h appears on both sides. We may therefore rewrite the dependent variable as $m_l/h^3\rho = m_l/m_0/c_0$, where c_0 is the nondimensional ratio accounting for the specific geometry of our geometrically similar weightlifters. This constant may be absorbed into a new function (see Rules of Thumb): $m_l/m_0 = F_2(\sigma_m/h\rho g)$.

In addition, we can use the definition of c_0 to eliminate the direct dependence on length scale as: $h^3 = m_0/c_0\rho \rightarrow h = 3\sqrt{m_0/c_0\rho}$. Whence,

$$m_1/m_0 = F_3(\sigma_m/(m_0/c_0\rho)^{1/3}\rho g) = F_3(\sigma_m/(m_0/c_0)^{1/3}\rho). \tag{14.23}$$

From the Buckingham Π-theorem we may write $\Pi_1 = f(\Pi_3)$, and so,

$$m_1/m_0 = m_0\sigma_m/m_0^{1/3}\rho^{2/3}g \rightarrow m_l = m_0^{2/3}\sigma_m/\rho^{2/3}g. \tag{14.24}$$

Santiago [6] goes on to develop a dimensional stress index that would allow the comparison of weightlifters across the different body weight classes of lifters.

Figure 14.5 presents data of the maximum weight deadlifted (to above the lifters head) by a range of male weightlifters (classed by body weight); the data is from an international competition held in 2019.We can clearly see that there is a steady

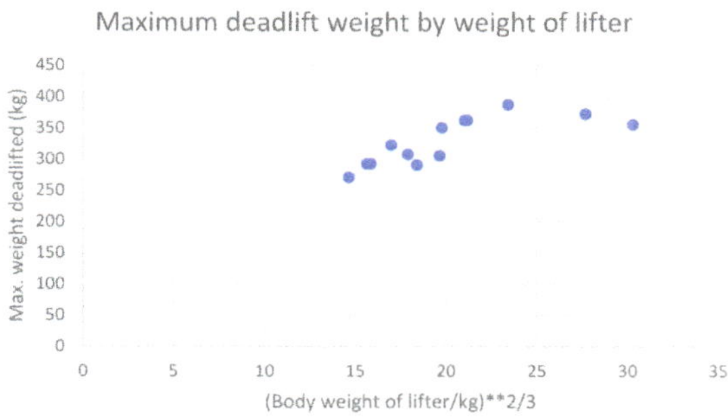

Figure 14.5 Results of the maximum weight deadlifted by many of the leaders of the various weight categories from the 49th World Open Men's Championship of the International Powerlifting Federation (Dubai, 18–23, November 2019). There are eight categories of weightlifter (defined by the lifter's body weight) in the men's competition. We have plotted m_l versus $(m_0)^{2/3}$. The data is taken from: International Powerlifting Federation IPF—International Powerlifting Federation IPF.

increase in weight lifted with respect to body weight of lifter, but this increase then reaches a plateau. The data has been scaled; in that we have plotted m_l against ${m_0}^{2/3}$, and the data below the plateau region (that is for the results of weightlifters with body mass < 120 kg) can clearly be seen to lie along a straight line passing through the origin (so any unidentified constants are small). The slope of this line is, $\sigma_m/\rho^{2/3}\ g$.

References

[1] Thompson D W 2014 *On Growth and Form* (Canto Classic edition) (Cambridge: Cambridge University Press)

[2] West G B, Brown J H and Enquist B J 1997 A general model for the origin of allometric scaling laws in biology *Science* **276** 122–6

[3] West G B and Brown J H 2005 The origin of allometric scaling laws in biology from genomes to ecosystems: towards a quantitative unifying theory of biological structure and organization *J. Exp. Biol.* **208** 1575–92

[4] Tennekes H (ed) 2009 *The Simple Science of Flight: From Insects to Jumbo Jets* (Cambridge MA: MIT Press)

[5] McNeill A R 1999 Walking and running *Math. Gaz.* **80** 488
McNeill A R 1991 *Dynamics of Dinosaurs and Other Extinct Giants* (New York: Columbia University Press)
Alexander R M 1983 A dynamic similarity hypothesis for the gaits of quadrupedal animals *J. Zool.* **201** 135–52

[6] Santiago J G 2019 *A First Course in Dimensional Analysis: Simplifying Complex Phenomena Using Physical Insight* (Cambridge, MA: The MIT Press)

[7] Fitzgerald R 2002 How fast could Tyrannosaurus rex run? *Phys. Today* **55** 18
Francis M R 2017 Deducing how dinosaurs moved *Phys. World* 2 February 2017

[8] McMahon T A 1971 Rowing: A similarity analysis *Science* **173** 349

Chapter 15

A miscellany of analyses by dimension

L'analyse mathématique … dans l'étude de tous les phénomènes; elle les interprète par le même langage, comme pour attester l'unité et la simplicité du plan de l'univers, et rendre encore plus manifeste cet ordre immuable qui préside à toutes les causes naturelles.

Jean-Baptiste-Joseph Fourier (1768–1830)

In this volume, I have tried to demonstrate the utility of the approximate method of problem solving known as dimensional analysis. We have seen how dimensional analysis is a technique that readily generates equations to explain complex phenomena in fields as disparate as fluid mechanics, molecular spectroscopy and biomechanics. However, this facility is also one of the great weaknesses of dimensional analysis. The equations derived by coupling the exponents from a power-series representation of the base dimensions involved in the phenomenon of interest are the simplest grouping of the variables involved. In addition, the equations generated by dimensional analysis do not contain numerical constants. Dimensional analysis is blind to numbers and constants such as π, which are essential if you are going to make comparisons between theory and experiment.

Indeed, dimensional analysis is a technique of analysis that, at first glance, one feels should be taught to all students, but upon reflection one realises that unless one has experience in handling quantities and in setting up, and then solving differential equations, the apparent simplicity of dimensional analysis may lead the unwary down some very deep rabbit holes. To bring this volume to a close, I include some of the dimensional analyses that could not be squeezed into other chapters. These are analyses of subjects slightly off the beaten-track, but they are still useful examples, and worth working through.

doi:10.1088/978-0-7503-3655-0ch15

15.1 Dimensional analysis of cooking

The physical concept of scaling is usually formulated in the language of dimensional analysis (see chapters 3 and 6). The central idea here is due to the French mathematician Egyptologist and politician, Jean-Baptiste-Joseph Fourier, who is most famous as a mathematician for the invention of Fourier analysis; introduced in his treatise *Théorie Analytique de la Chaleur*, first published in Paris in 1822 [1]. It was Fourier who first realised that every physical quantity '*has one dimension proper to itself, and that the terms of one and the same equation could not be compared, if they had not the same exponent of dimension.*' He goes on: '*We have introduced this consideration … to verify the analysis … it is the equivalent of the fundamental lemmas which the Greeks have left us without proof.*' However, as we have seen, it was only much later that physicists began to use the '*method of dimensions*' to solve complex problems.

As an exercise in dimensional analysis, let us consider the problem of cooking a turkey (or any piece of meat). We place the bird in an oven with temperature T_0, assumed constant. The bird itself has mass M and density ρ. The problem we wish to solve is: how long (t) must the bird remain in the oven for its central temperature to reach some specified (safe to eat) value, T_c?

The physics of the situation is governed by the diffusion equation due to Fourier:

$$\frac{\partial T}{\partial t} = \kappa \nabla^2 \mathrm{T}, \tag{15.1}$$

where $T(x,y,z,t)$ is the temperature at any point in the turkey (labelled by x,y,z) at time t. The constant κ is the thermal diffusion coefficient, and depends on the material being heated. Thus, κ might be different for turkey and beef, but all turkeys will share essentially the same κ. One way to view thermal diffusivity is as the ratio of the time derivative of temperature to its curvature, quantifying the rate at which temperature concavity is smoothed out—it is the measure of thermal inertia. In a substance with high thermal diffusivity, heat moves rapidly through it because the substance conducts heat quickly relative to its volumetric heat capacity or thermal bulk. From looking at the dimensions in this equation we can determine the units of κ. The operator ∇^2 contains second derivatives with respect to x, y and z, so it has dimensions 1 length^{-2}, or L^{-2}. Hence κ has dimensions, $[\kappa] = \mathsf{L}^2\mathsf{T}^{-1}$, where T is time (dimension, not variable); in the SI system, the unit would be m^2 s^{-1} (values range: $\kappa = 0.12 \times 10^{-6}$ m^2 s^{-1} for raw beef-steak to 111×10^{-6} m^2 s^{-1} for copper).

The dimensions of the quantities involved in this problem imply a general structure of the result we seek. Let ΔT_m be the difference in temperature between the raw (but defrosted) meat and the hot oven, and ΔT_c be the temperature between the temperature of the cooked meat and the oven (we assume the oven temperature is kept constant). These two temperature differences will have a significant influence on the time of cooking. A measure of the factor representing the difference in cooking times for different foodstuffs is the coefficient of heat conduction—the thermal diffusivity, κ. Thus, we can say that the cooking time, t, of our turkey will be a function of the form,

$$t = f(\Delta T_m, \Delta T_c, \kappa, l), \qquad (15.2)$$

where l is a measure of the size of the turkey. First consider the dimensions of the independent variables. The temperature variables, ΔT_m and ΔT_c, represent energy per volume, and therefore have the dimensions $ML^2T^{-2}/L^3 = ML^{-1}T^{-2}$. The thermal conductivity of the raw meat, or of anything else, is defined as the amount of energy crossing unit cross-sectional area per second divided by the gradient perpendicular to the area; that is:

$$\kappa = ML^2T^{-2}.\ L^{-2}T^{-1}/ML^{-1}T^{-2}.\ L^{-1} = L^2T^{-1}. \qquad (15.3)$$

The dimensions of the variable involved in this cooking procedure (cooking is just like experimentation, except you eat the product) are given in table 15.1.

Any product of the variables must be of the form $(\Delta T_m)^a\ (\Delta T_c)^b\ \kappa^c\ l^d\ t^e$, and hence have dimensions $(ML^{-1}T^{-2})^a(ML^{-1}T^{-2})^b(L^2T^{-1})^c\ (L)^d\ (T)^e$.

From this set of dimensions, we may derive the dimensional equation,

$$[t] = T = M^{a+b}L^{-a-b+2c+d}T^{-2a-2b-c+e}. \qquad (15.4)$$

As there are five unknown exponents, and only three equations (only three base dimensions are involved) so a unique solution is not possible; the solutions of equation (15.4) are: $a = -b$, $c = e$ and $d = -2e$, where b and e are arbitrary constants. If we set $b = 1$, $e = 0$, we obtain $a = -1$, $c= 0$ and $d = 0$; likewise, $b = 0$, $e = 1$ gives $a = 0$, $c = 1$ and $d = -2$. These independent solutions yield the complete set of dimensionless products (alternatively, we could have chosen repeat variables and used the Π-theorem mechanism to discover the two nondimensional groups of variables): $\Pi_1 = \Delta T_m^{-1}\ \Delta T_c$, and $\Pi_2 = \kappa t/l^2$.

From the Buckingham theorem, we obtain:

$$f(\Pi_1, \Pi_2) = 0, \text{ or } t = (l^2/\kappa)f(\Delta T_m^{-1}\Delta T_c). \qquad (15.5)$$

Let us assume turkeys are all geometrically similar, with volume (V), $V \propto l^3$. If we further assume that the turkey has a constant density (this is not such a robust assumption), then because mass and weight are given by density times volume, and

Table 15.1. Variables for cooking.

Variable	Dimension	Description
ΔT_m	$ML^{-1}T^{-2}$	+Temperature as energy/volume; that is, Energy/L^3 = $ML^2T^{-2}/L^3 = ML^{-1}\ T^{-2}$
ΔT_c	$ML^{-1}T^{-2}$	+Temperature as energy/volume; that is, Energy/L^3 = $ML^2T^{-2}/L^3 = ML^{-1}\ T^{-2}$
κ	$L^2\ T^{-1}$	Thermal conductivity
l	L	Size of turkey
t	T	Cooking time

+The art of solving problems with dimensional analysis is to be imaginative, and versatile in the way quantities are expressed.

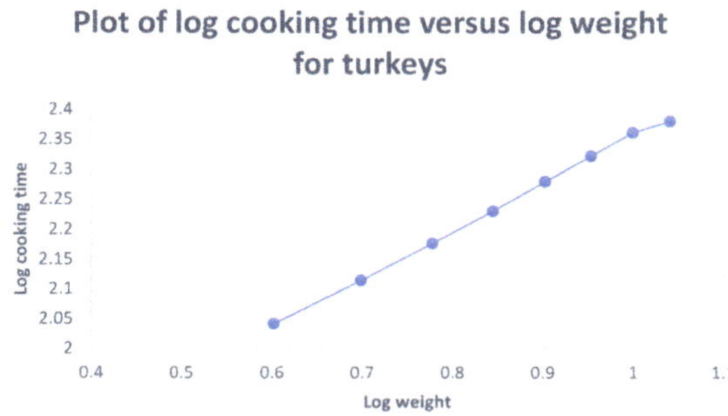

Figure 15.1. A log–log plot of cooking time versus weight of turkey cooked.

volume is proportional to l^3, we have weight $\propto l^3$. Moreover, if we set the oven to constant baking temperature, and specify that the initial temperature of the raw meat be near room temperature, then $\Delta T_{\text{m}}^{-1}\ \Delta T_{\text{c}}$ is a dimensionless constant. Combining these results, we may write for the cooking time, t, $t \propto \text{weight}^{2/3}$.

There is a standard procedure for testing this kind of power law relation. If we take the log of both sides of this relation, we obtain $\log(t) = (2/3)\log(\text{weight}) + C$; so that if we plot the log of the cooking time versus the log of the weight, we should see a straight line with slope 2/3, the power we are trying to verify. Figure 15.1 displays some data taken from a Google search of cooking times of different-sized turkeys (4 kg–11 kg). The plot is indeed linear, but with slope of about 3/4, rather than 2/3. Interestingly, the same slope is obtained from a log–log plot of cooking times for different-sized turkeys covered in aluminium cooking foil. For the foil-covered turkeys, the cooking time at any weight is longer, as the metal foil keeps the moisture in (increasing succulence and eliminating the need for basting), and so limits the rapidity of the temperature rise.

15.2 Black holes

As we are all aware, a black hole is a region of space-time where gravity is so intense that nothing—no particles, nor even electromagnetic radiation may escape its force of attraction. General relativity predicts that a sufficiently compact mass can deform space-time to form a black hole. The boundary of the region within which escape is impossible from the gravitational attraction is called the event horizon. Although the event horizon has an enormous effect on the fate and circumstances of an object crossing it; according to general relativity it has no locally detectable features. In many ways, a black hole acts like an ideal black-body, as it reflects no light.

Black holes of stellar mass are expected to form when very massive stars collapse at the end of their life-cycle. After a black hole has formed, it can continue to grow by absorbing mass from its surroundings. By absorbing other stars and merging with other black holes, supermassive black holes of millions of solar masses may form. The presence of a black hole can be inferred through its interaction with other matter, and with electromagnetic radiation such as visible light. If there are other stars orbiting a black hole, their orbits can be used to determine the black hole's mass and location. In this way, astronomers have identified numerous stellar black hole candidates in binary systems. But on 10 April 2019, the first direct image of a black hole and its vicinity was published, following observations made by the Event Horizon Telescope in 2017 of the supermassive black hole in the centre of the Galaxy, Messier 87 (see figure 15.2).

Black holes may seem to be exotic, but the laws of physics must apply to them as they do here on Earth; that is, the same constants of Nature that we use here to define phenomena, may be used to define the strange phenomena that are black holes. Let us see if we can estimate the size of the black hole; i.e. light-trapping region, based on its mass and some dimensional analysis. Since black holes are heavy, when considering which variables we need to consider we must include G (Newton's universal gravitation constant), and since light is involved, we should also include c.

We assume the radius, R, of the black hole is related to its mass m by

$$R \propto (m)^a (G)^b (c)^d. \tag{15.6}$$

Taking dimensions (see table 15.2) we get:

$$\mathsf{L} = (\mathsf{M})^a(\mathsf{L}^3\mathsf{M}^{-1}\mathsf{T}^{-2})^b(\mathsf{L}\mathsf{T}^{-1})^d = \mathsf{M}^{a-b}\mathsf{L}^{3b+d}\mathsf{T}^{-(2b+d)}. \tag{15.7}$$

The solutions are: $a = 1$, $b = 1$ and $d = -2$. The size of a black hole is thus related to its mass by

$$R \sim Gm/c^2. \tag{15.8}$$

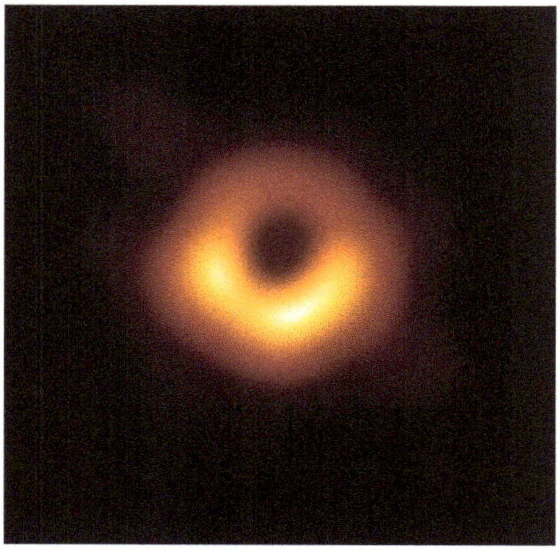

Figure 15.2. The supermassive black hole at the core of supergiant elliptical galaxy Messier 87, with a mass about 7 billion times that of the Sun. What we see are the crescent-shaped emission rings, as matter is excited and emitting radiation before it falls into the central shadow. Picture credit: EHT Collaboration.

As is usual, asking a specific question will generate a specific answer—but one devoid of numerical constants; the radius of a spherical black hole of mass, m, is $R_s = 2Gm/c^2$. This is called the Schwarzschild radius. The escape velocity v_{esc} for a particle of mass, $m_{particle}$, in the gravity potential, generated by a large mass $M_{black\ hole}$ is defined as the velocity needed to escape the gravity well (depth, E_{grav}), to get infinitely far away with zero kinetic energy (E_{kin}) left over. This implies that

$$E_{kin} + E_{grav} = ½\ m_{particle}\ v_{esc}{}^2 - \frac{GM}{R}\ m_{particle} = 0. \quad (15.9)$$

If we set $v_{esc} = c$ and solve for R:

$$R = G\ M_{blackhole}\ m_{particle}/½\ m_{particle}\ c^2 = 2GM_{blackhole}/c^2. \quad (15.10)$$

The escape velocity equals the speed of light $v_{esc} = c$ at the Schwarzschild radius.

Consider a box of hot gas. From a macroscopic perspective, it has only a few coarse features—volume and energy. But microscopically, there are, of order, 10^{23} atoms or molecules moving about rapidly, and continually colliding with each other. The black hole is perhaps similar; it has a volume $V \sim R^3$ and energy $E = mc^2$. For the moment, we will simply hypothesise that, like the box of gas, the black hole is a thermodynamic system. This means the Boltzmann constant, k_B, will be involved in any discussion of the thermodynamics of a black hole. If a black hole is a thermodynamic system, we can imagine it has a temperature. But what would this mean? On Earth, hot objects radiate. Since black holes form in the vacuum of space, radiation would be the only way for us to tell if they have a temperature. We also need to consider the introduction of the Planck constant, h, into any analysis of black holes.

We will now consider the temperature of a black hole (T_{BH}). We will treat R in equation (15.8) as a fixed system size, and ignore G for this analysis. We hypothesise that

$$T_{BH} \propto (R)^a (k_B)^b (h)^d (c)^e. \quad (15.11)$$

These variables are defined in table 15.2.

The dimensional equation is

$$\theta = (\mathrm{L})^a (\mathrm{ML^2T^{-2}\theta^{-1}})^b (\mathrm{ML^2T^{-1}})^d (\mathrm{LT^{-1}})^e = \mathrm{M}^{b+d}\mathrm{L}^{a+2b+2d+e}\mathrm{T}^{-(2b+d+e)}\theta^{-b}. \quad (15.12)$$

Table 15.2. Variables for a black hole.

Variable	Dimension	Description
G	$\mathrm{M^{-1}L^3T^{-2}}$	Universal gravitational constant
m	M	Mass of black hole
T_{BH}	θ	Temperature of black hole
R	L	Radius of black hole
k_B	$\mathrm{ML^2T^{-2}\ \theta^{-1}}$	Boltzmann constant
h	$\mathrm{ML^2T^{-1}}$	Planck constant
c	$\mathrm{LT^{-1}}$	Speed of light
A	$\mathrm{L^2}$	Surface area

The solutions are: $a = b = -1$ and $d = e = 1$. This leads to an estimate of the temperature of a black hole, $T_{\mathrm{BH}} \sim K\ h.c./k_{\mathrm{B}}R$, where K is the constant of proportionality. From equation (15.8),

$$T_{\mathrm{BH}} \sim K\ hc^3/k_{\mathrm{B}}Gm. \tag{15.13}$$

This agrees well with Hawking's calculation of the Hawking temperature; that is, up to a numerical constant that is invisible to dimensional analysis. Hawking discovered that $T_{\mathrm{BH}} = hc^3/8\pi k_{\mathrm{B}}Gm$. This result was a complete shock to Hawking, and the rest of the physics community; to discover that black holes could radiate energy. The Stefan-Boltzmann law (see section 10.1) tells us that a black hole will likely be a perfect emitter of radiation that will lose radiant power (P) at a rate

$$P = \sigma A T^4. \tag{15.14}$$

The energy of a system of N hot particles is $E \sim Nk_{\mathrm{B}}T$. The energy of a black hole is $E = mc^2$. If we assume the thermodynamic nature of black holes, we should be able to equate these two forms of energy. Since we have an expression for the temperature, equation (15.13), we can also estimate N, the number of particles in the black hole, using equation (15.8):

$$N \sim E/k_{\mathrm{B}}T \sim Gm\ mc^2/\hbar c^3 \sim (Gm/c^2)^2\ c^3\ \ G^{-1}\hbar \sim R^2c^3/G\hbar. \tag{15.15}$$

Equation (15.15) tells us that the effective number of particles N in the black hole scales like the area, $A \sim R^2$, rather than the volume $V \sim R^3$. This suggests the thermodynamic system underlying the black hole is a surface phenomenon. In other words, the maximum number of particles you can fit into a region V, and still describe using gravity, scales as the surface area of the region. Since this is similar to the way a 3-D image emerges from a 2-D grating in optical holograms, this property of gravity is called the holographic principle.

Usually, the dimension N describes the amount of something for a system consisting of many particles. Here we will use N for the measurement of the particles on the surface of the black hole. This means that $[A] = \mathsf{L}^2\mathsf{N}$; whereas for an atomic-size patch of the surface, a, $[a] = \mathsf{L}^2$. The power emitted by the black hole is the sum of power emitted from each atomic patch of its surface. Thus, we expect the radiated power, P, to also pick up a factor N, so $[P] = \frac{[\text{Energy}]}{[\text{Time}]} = \mathsf{M}\ \mathsf{N}\ \mathsf{L}^2\ \mathsf{T}^{-3}$. We are saying, A and P scale with the system size. For our radiating black hole, we assume

$$P \propto (h)^a (k_{\mathrm{B}})^b (c)^d (A)^f (\theta)^g. \tag{15.16}$$

We have five unknown quantities, and five base dimensions (M, T, L, θ, N) so we should expect to be able to determine the five exponents uniquely. On the LHS, we have $[P] = \mathsf{ML^2NT^{-3}}$, and on the RHS, we have:

$$\begin{aligned}&(\mathsf{ML^2T^{-1}})^a(\mathsf{ML^2T^{-2}}\theta^{-1})^b(\mathsf{LT^{-1}})^d(\mathsf{L^2\ N})^f(\theta)^g\\&= \mathsf{M}^{a+b}\mathsf{L}^{2a+2b+d+2f}\mathsf{T}^{-a-2b-d}\theta^{g-b}\mathsf{N}^f.\end{aligned} \tag{15.17}$$

The results are: $a = -3$, $b = 4$, $d = -2$, $g = 4$ and $f = 1$. So we find

$$P = K(k_B{}^4/h^3c^2)AT^4, \tag{15.18}$$

where K is the constant of proportionality, which will include the many terms lost in the dimensional analysis.

Using the total energy of the black hole to be $E = mc^2$, we may estimate the lifetime, t_{evap} of the black hole in terms of the black hole mass m; that is, how long will it take a black hole to evaporate away due to radiating thermal energy. The rate of power loss is $P = \sigma AT^4$. We suppose an evaporation time-scale $t_{evap} \sim E/P$, since P is the rate the black hole is radiating away its stored energy. We can make the dependence on m explicit by expanding out our definitions using $A \sim R^2$, the definition of the Stephan-Boltzmann constant, and equations (15.8) and (15.13):

$$t_{evap} \sim \frac{E}{P} \sim mc^2/\sigma AT^4$$

$$\sim mc^2\, k_B{}^2R^4/\sigma R^2h^4c^4$$

$$\sim mk_B{}^2h^3c^2G^2m^2/h^4c^2k_B{}^2c^4 = G^2m^3/hc^4. \tag{15.19}$$

If you put numbers into this equation, you will find that it will take a long time for a black hole to evaporate away. A black hole of the mass of our Sun would take about 10^{55} years to evaporate away. But the Universe has only been in existence for about 10^{10} years. A black hole of one solar mass has a temperature of only 60 nanokelvins, and such a black hole would, in any case, absorb more cosmic microwave background radiation than it emits. A black hole of 4.5×10^{22} kg (about the mass of the Moon, or about 133 μm across) would be in equilibrium at 2.7 K, absorbing as much radiation as it emits.

15.3 The Aeolian harp

And that simplest lute,
Placed length-ways in that clasping casement, hark!
How by the desultory breeze caressed,...
... And now, it's strings
Boldlier swept, the loud sequacious notes
Over delicious surges sink and rise,
Such a soft floating witchery of sound
As twilight Elfins make, when they at eve
Voyage on gentle gales from Fairy-Land,
From *The Aeolian Harp* by Samuel Taylor Coleridge (1795)

The Aeolian, or wind harp, is a musical instrument played by the wind. Named for Aeolus, the Greek god of the wind, the traditional Aeolian harp is essentially a wooden box including a sounding board, with strings stretched lengthwise across two bridges. It is often placed in a slightly opened window where the wind can blow across the strings to generate sounds. The strings can be made of different materials (and thicknesses), and all be tuned to the same pitch, or identical strings can be tuned to different pitches. The quality of sound depends on many factors, including the lengths, gauges, and types of strings, the character of the wind, and the material of the resonating body. Metal-framed instruments with no sound board produce music very different from that produced by wind harps with wooden sound boxes and sound boards. There is no percussive aspect to the sound like that produced by a wind chime; rather crescendos and decrescendos of harmonic frequencies are played in rhythm to the winds. (An Aeolian harp on the Irish coast may be heard at [2].)

The Aeolian harp—already known in the ancient world—was first described in modern times by Athanasius Kircher (1602–1680) in his *Phonurgia Nova* (1673). These instruments became popular during the Romantic era, and Aeolian harps are still hand-crafted today. Some are now made in the form of monumental metal sound-sculptures located on the roof of a building, or a windy hilltop. Although the polymath Athanasius Kircher investigated the mechanism of the aeolian harp, it was not until the end of the 19th Century that a first attempt was made to explain the physics of the device [3]. Then in 1915, Lord Rayleigh (see figure 15.3) used dimensional analysis to give the full explanation [4].

Early research defined a useful relationship between the tone frequency (f), air speed (u) over the wire or string, and the string diameter (d):

$$\mathrm{St} = \frac{fd}{u}. \tag{15.20}$$

Here, St is a dimensionless parameter named after Vincenc Strouhal (1850–1922), a Czech physicist who in 1878, first systematically experimented with wires experiencing vortex shedding and singing in the wind [3]. The Strouhal number is important when analysing unsteady, oscillating flow problems. It represents a measure of the ratio of the inertial forces due to the unsteadiness of the flow, or local acceleration to the inertial forces due to changes in velocity from one point to another in the flow field; for example, the vortices in a fast-flowing river observed behind the pillars, or support of a bridge. This relationship is readily arrived at by dimensional analysis. If $f = (d)(u)$, we may write in dimensional terms $[f] = \mathsf{T}^{-1} = [d][u] = \mathsf{L.LT}^{-1}$; a nondimensional combination of these three variables gives $\frac{fd}{u}$. In certain circumstances this characteristic length, d, is the amplitude of oscillation. This selection of characteristic length can be used to present a distinction between the Strouhal number and reduced frequency: $\mathrm{St} = \frac{ka}{\pi c}$, where k is the reduced frequency, and a is the amplitude of the heaving oscillation.

For large Strouhal numbers (of order, 1), viscosity dominates fluid flow, resulting in a collective oscillating movement of the fluid 'plug'. For low Strouhal numbers (order of 10^{-4} and below), the high-speed, quasi-steady-state portion of the movement dominates the oscillation. Oscillation at intermediate Strouhal numbers is characterized by the build-up and rapidly subsequent shedding of vortices [5].

Figure 15.3. John William Strutt, 3rd Baron Rayleigh (1842–1919), who not only explained the Aeolian harp, but invented the method of dimensional analysis. Reproduced from https://en.wikipedia.org/wiki/John_William_Strutt,_3rd_Baron_Rayleigh. Image stated to be in the public domain. It is included within this article on that basis.

In swimming or flying animals, the Strouhal number is defined as $\mathrm{St} = \frac{f}{u} A$, where f is the oscillation frequency (tail-beat, wing-flapping, etc), u is the air flow rate, and A is the peak-to-peak oscillation amplitude.

As with a great many concepts in fluid mechanics, Strouhal numbers play a significant role in the natural world. In animal flight or swimming, propulsive efficiency is high over a narrow-range of Strouhal numbers, generally peaking in the $0.2 < \mathrm{St} < 0.4$ range. This range is used in the swimming of dolphins, sharks and bony fish, and in the cruising flight of birds, bats and insects. However, in other forms of flight other values are found. Intuitively, the ratio measures the steepness of the strokes, viewed from the side (e.g., assuming movement through a stationary fluid), with f being the stroke frequency, A is the amplitude, so the numerator fA is half the vertical speed of the wing tip, while the denominator u is the horizontal speed. Thus, a graph of the wing tip forms an approximate sinusoid with aspect (maximal slope) twice the Strouhal number.

In Strouhal's initial experiments on Aeolian harps, a vertical wire or rod attached to a suitable frame was caused to revolve with uniform velocity about a parallel axis. The pitch of the Aeolian tone generated by the relative motion of the wire and of the air was found to be independent of the length and of the tension of the wire, but to vary with the diameter and with the speed of the motion. Within certain limits, the relation between the frequency of vibration (f) and these observations were found to be well-represented by $f = 185\,\frac{u}{d}$.

When the velocity is such that the Aeolian tone coincides with one of the natural, or proper resonances of the taut wire, supported so as to be capable of free independent vibration, the generated sound is greatly reinforced by resonant enhancement. Strouhal further showed that with a given diameter, and a given velocity, an increase in temperature resulted in a fall in pitch.

If the compressibility of the air is omitted from the dimensional analysis, we may regard f as a function of: the relative velocity u (LT^{-1}), the diameter of the wire, d (L), and υ the kinematic coefficient of viscosity (L^2T^{-1}). In this case f is of the form (four variables, but only two dimensions; so, there are only two nondimensional groups),

$$f = \frac{u}{d}\mathrm{F}(\upsilon/\mathrm{ud}), \tag{15.21}$$

where F represents an arbitrary function; and there is dynamical similarity if $\upsilon \propto ud$. In measurements in air at one temperature, υ is constant; and if d varies inversely as u, fd/u should be constant—as observed by Strouhal. Again, if the temperature rises, υ increases, and in order to accord with Strouhal's observations, we must suppose that the function F diminishes with increasing argument.

In a *tour de force* of dimensional analysis, Lord Rayleigh [4] demonstrated the range of physical conditions; including the nature of the string or cord that influenced the frequency of the generated tone. Lord Rayleigh thought that the dimensionless quantity υ/ud was always small, so he expanded the expression for f in a MacLaurin series. He took $f(x) = \mathrm{a} + bx + cx^2$, and was able to write from equation (15.21),

$$f = a\frac{u}{d} + b(\upsilon/d^2) + c(\upsilon^2/ud^3)\ldots \tag{15.22}$$

(each term having the dimensions of frequency, T^{-1}). If the third term on the RHS is neglected, there is a linear relationship between f and u. The coefficient, b, is found to be negative, which is required to explain the observed temperature dependence of f. Further,

$$\mathrm{d}\frac{df}{du} = \mathrm{a} - \{c\upsilon^2/u^2\mathrm{d}^2\}; \tag{15.23}$$

so $d\frac{df}{du}$ is very nearly constant—the result given by Strouhal [3].

Lord Rayleigh fitted the measurements of Strouhal to the model (in c.g.s. units),

$$\frac{fd}{u} = 0.195\left(1 - \frac{3.02}{ud}\right). \tag{15.24}$$

Although, the agreement is good, Lord Rayleigh observed that a change of wire introduces greater discrepancies than a change in u; a circumstance which may possibly be attributed to alterations in the character of the surface. The simple form of equation (15.20) assumes that the wires are smooth, or else that the roughness is in proportion to *d*, so as to secure geometrical similarity. (For the importance of surface smoothness, see the discussion on golf-balls, section 7.3.)

15.4 The final frontier of dimensional analysis: the Drake equation

As dimensional analysis is a technique ideally suited to providing an insight into hypothesised functions and equations concerning phenomena that are difficult to examine in detail, let us consider an equation about which it is impossible—at present—to have a direct observation. The Drake equation is a probabilistic argument used to estimate the number of active, communicative extraterrestrial civilizations in our galaxy.

The equation in question, was written down in 1961 by astronomer Frank Drake (born 1930), not for purposes of quantifying the number of civilizations, but as a way to stimulate dialogue at the first scientific meeting on the Search for ExtraTerrestrial Intelligence (SETI)[1]. The equation summarizes the main concepts which scientists must contemplate when considering the question of other life-forms capable of radio-communication. The equation is, perhaps, best thought of as an approximation, or a back-of-the-envelope calculation rather than as a serious attempt to determine a precise number. Criticism related to the Drake equation focuses not on the equation itself, but on the fact that the estimated values for several of its variables are highly conjectural; the combined multiplicative effect being that the uncertainty associated with any derived value is so large that the equation cannot be used to draw firm conclusions. And for these reasons, we include it at the end of a volume dedicated to dimensional analysis and the use of intuitive thinking in solving problems in physics.

Thinking about extraterrestrial life goes back at least as far as Greek antiquity. The main argument for the existence of life-forms and intelligent beings beyond Earth was most clearly formulated by Metrodorus of Chios, a philosopher of the school of Democritus, in the 4th Century BCE: '*To consider the Earth as the only populated world in infinite space is as absurd as to assert that in an entire field sown with millet only one grain will grow*.' The idea of space being infinite, with an infinite number of atoms populating it and composing its various objects, was a key ingredient of the atomistic philosophy of Leucippus, Democritus and Epicurus, and was re-asserted in more recent times by Giordano Bruno (1548–1600).

The argument of Metrodorus of Chios is invoked today by proponents of extraterrestrial intelligence (ETI) essentially unaltered, although the concept of infinity is no longer used, because it is difficult to handle and it may lead to paradoxes; for example, in an infinite universe, everything—including ourselves—could exist in an infinite number of copies. So, it has been replaced by the phrase 'other intelligences in our Milky Way galaxy', which contains about 100 billion stars. This number is considered by some—mostly astronomers—to be large enough as to make Metrodorus' argument applicable to the Milky Way; while others—

[1] Although given the consequences of contacting extraterrestrial civilizations as described in the splendid science fiction novels of the Chinese writer, Liu Cixin: *The Three-Body Problem*, *The Dark Forest* and *Death's End*, perhaps we are better off not looking for our neighbours out in space.

especially biologists—are not impressed by that number and remain skeptical concerning ETI.

In its original formulation, the Drake equation is:

$$N = R^* f_p n_e f_l f_i f_T L, \tag{15.25}$$

where R^* is the rate of star formation in our galaxy (i.e. number of stars per unit time), f_p is the fraction of stars with planetary systems, n_e is the average number of planets around each star, f_l is the fraction of planets where life developed, f_i is the fraction of planets where intelligent life developed, and f_T is the fraction of planets with technological civilizations. N and L are intimately connected: if N is the number of radio-communicating civilizations—as in the original formulation by Drake—then L is the average duration of the radio-communication phase of such civilizations (and not their total lifetime, as is sometimes incorrectly stated).

As can be seen, the equation is a product of probabilities; each variable taking a value between one and zero, but nevertheless viewed from the point of terrestrial biology and the history of humanity. Although never explicitly stated, the Drake formula corresponds to the equilibrium solution (a steady-state) of an equation similar to the well-known equation of radioactivity $\frac{dN}{dt} = -N/L$, where N is the number of radioactive nuclei in a sample, and L is their half-life. In the steady-state, where the production rate, P, is equal to the decay rate, $D = \frac{dN}{dt} = -N/L$, one has $N = PL$. In a similar vein, the product of all the terms of the Drake formula, except L can be interpreted as the production rate P of radio-communicating (or technological, or space-faring) civilizations in the Galaxy. The trouble is that in the Drake equation, time is a fluid concept.

As can be seen in this volume, even the most obscure of problems in dimensional analysis affords ample opportunity to examine physics and the nature of equations and functions. A recent study [6] has calculated that there could be about 36 active civilizations, capable of communicating in our galaxy. However, due to time and distance, we may never actually know if they exist, or ever existed.

The Nottingham group [6] developed what they call the Astrobiological Copernican Principle[2] to establish weak and strong limits on life in the Galaxy (refining the definitions and limits of the original Drake equation). Their version of the Drake equation includes the history of star formation in our galaxy and the ages of stars, the metal content of the stars, and the likelihood of stars hosting Earth-like planets in their habitable zones where life could form. The habitable zone is the right distance from a star, not too hot or too cold, where liquid water and life as we know it may be possible on the surface of a planet. The authors point out that of these factors, habitable zones are critical, but that orbiting a quiet, stable star for billions of years may be more critical. The Astrobiological Copernican Strong limit is that life must form between 4.5 to 5.5 billion years after the formation of the planet, as on

[2] It is called the Astrobiological Copernican Principle because it makes the assumption that our existence is not special. That is, if the conditions in which intelligent life on Earth developed somewhere else in the Galaxy—then intelligent life would develop there in a similar manner.

Earth, while the weak limit states that a planet takes, at least, four billion years to form life, but it can form any time after that. This assumes that life forms the way it does on Earth, which is our only understanding at the moment.

References

[1] Fourier J B J 2009 *The Analytical Theory of Heat* (originally published 1822) (Cambridge: Cambridge University Press)

[2] An aeolian harp many be heard at: Wind Harp—Aeolian Harp on the Irish coast—YouTube https://youtu.be/rmP5XaNYlkI

[3] Strouhal V 1878 *Über eine Besondere Art der Tonerregung* (On an unusual sort of sound excitation) *Ann. Phys. Chem.* 3rd series **5** 216–51

[4] Lord Rayleigh 1915 Æolian tones *Lond. Edinb. Dublin Philos. Mag. J. Sci.* [Sixth Series] **29** 433–44

[5] A full description of the physics of aeolian harps may be found in; Powell A 1990 Some aspects of aeroacoustics from Rayleigh until today *J. Vib. Acoust.* **112** 145–59

[6] Westby T and Conselice C J 2020 The astrobiological Copernican weak and strong limits for intelligent life *Astrophys. J.* **896** 58

www.ingramcontent.com/pod-product-compliance
Ingram Content Group UK Ltd.
Pitfield, Milton Keynes, MK11 3LW, UK
UKHW051857100926
8556IPUK00005B/658

9 780750 33653